Springer Theses

Recognizing Outstanding Ph.D. Research

Aims and Scope

The series "Springer Theses" brings together a selection of the very best Ph.D. theses from around the world and across the physical sciences. Nominated and endorsed by two recognized specialists, each published volume has been selected for its scientific excellence and the high impact of its contents for the pertinent field of research. For greater accessibility to non-specialists, the published versions include an extended introduction, as well as a foreword by the student's supervisor explaining the special relevance of the work for the field. As a whole, the series will provide a valuable resource both for newcomers to the research fields described, and for other scientists seeking detailed background information on special questions. Finally, it provides an accredited documentation of the valuable contributions made by today's younger generation of scientists.

Theses are accepted into the series by invited nomination only and must fulfill all of the following criteria

- They must be written in good English.
- The topic should fall within the confines of Chemistry, Physics, Earth Sciences, Engineering and related interdisciplinary fields such as Materials, Nanoscience, Chemical Engineering, Complex Systems and Biophysics.
- The work reported in the thesis must represent a significant scientific advance.
- If the thesis includes previously published material, permission to reproduce this must be gained from the respective copyright holder.
- They must have been examined and passed during the 12 months prior to nomination.
- Each thesis should include a foreword by the supervisor outlining the significance of its content.
- The theses should have a clearly defined structure including an introduction accessible to scientists not expert in that particular field.

More information about this series at http://www.springer.com/series/8790

Kento Masuda

Exploring the Architecture of Transiting Exoplanetary Systems with High-Precision Photometry

Doctoral Thesis accepted by
the University of Tokyo, Tokyo, Japan

Author
Dr. Kento Masuda
Department of Astrophysical Sciences
Princeton University
Princeton, NJ
USA

Supervisor
Prof. Yasushi Suto
Department of Physics
The University of Tokyo
Tokyo
Japan

ISSN 2190-5053   ISSN 2190-5061 (electronic)
Springer Theses
ISBN 978-981-10-8452-2   ISBN 978-981-10-8453-9 (eBook)
https://doi.org/10.1007/978-981-10-8453-9

Library of Congress Control Number: 2018933005

© Springer Nature Singapore Pte Ltd. 2018
This work is subject to copyright. All rights are reserved by the Publisher, whether the whole or part of the material is concerned, specifically the rights of translation, reprinting, reuse of illustrations, recitation, broadcasting, reproduction on microfilms or in any other physical way, and transmission or information storage and retrieval, electronic adaptation, computer software, or by similar or dissimilar methodology now known or hereafter developed.
The use of general descriptive names, registered names, trademarks, service marks, etc. in this publication does not imply, even in the absence of a specific statement, that such names are exempt from the relevant protective laws and regulations and therefore free for general use.
The publisher, the authors and the editors are safe to assume that the advice and information in this book are believed to be true and accurate at the date of publication. Neither the publisher nor the authors or the editors give a warranty, express or implied, with respect to the material contained herein or for any errors or omissions that may have been made. The publisher remains neutral with regard to jurisdictional claims in published maps and institutional affiliations.

Printed on acid-free paper

This Springer imprint is published by Springer Nature
The registered company is Springer Nature Singapore Pte Ltd.
The registered company address is: 152 Beach Road, #21-01/04 Gateway East, Singapore 189721, Singapore

Supervisor's Foreword

Since the first discovery of an exoplanet orbiting around a Sun-like star, exoplanets have been established as one of the most fascinating research areas in astronomy. In particular, the Kepler mission has told us numerous surprising diversities of exoplanetary systems, which are supposed to provide a variety of key clues to their origin and evolution over the cosmological timescales.

In this thesis, the author presents several different approaches to the problem. After a brief overview of the observed diversity of exoplanetary systems in Chap. 1, he reviews the observational methodology to determine the obliquity of host stars and describes possible channels to spin–orbit misalignment in Chaps. 2 and 3. The following chapters correspond to his original published results: first determination of the true (not projected) spin–orbit angle for the transiting multi-planetary system of a main-sequence host star, Kepler-25, and that for HAT-P-7 from a combined analysis of asteroseismology, transit photometry, and the Rossiter–McLaughlin effect (Chap. 4). He was able to derive beautifully the spin–orbit misalignments for Kepler-13Ab and HAT-P-7b from the precise modeling of gravity darkening effect (Chap. 5). Finally, he successfully determined the architecture of the hierarchical triple system KIC 6543674 from the Kepler photometry alone (Chap. 6).

The methodologies that he described in each chapter will find broader applications in future data, and promise to provide fundamental contribution to understanding formation and evolution of exoplanetary systems.

I am confident that this book serves as a good introduction for readers who are interested in accurate modeling and characterization of exoplanetary systems.

Tokyo, Japan
January 2018

Yasushi Suto

Parts of this thesis have been published in the following journal articles:

Chapter 4—Benomar, O., Masuda, K., Shibahashi, H., Suto, Y. "Determination of three-dimensional spin–orbit angle with joint analysis of asteroseismology, transit lightcurve, and the Rossiter–McLaughlin effect: Cases of HAT-P-7 and Kepler-25," *Publications of the Astronomical Society of Japan*, 66 (2014) 94 (21pp.) Published by Oxford University Press on behalf of the Astronomical Society of Japan.

Chapter 5—Masuda, K. "Spin–Orbit Angles of Kepler-13Ab and HAT-P-7b from Gravity-darkened Transit Light Curves," *The Astrophysical Journal*, 805 (2015) 28 (14pp.) Published by IOP Publishing for the American Astronomical Society.

Chapter 6—Masuda, K., Uehara, S., Kawahara, H. "Absolute Dimensions of a Flat Hierarchical Triple System KIC 6543674 from the Kepler Photometry," *The Astrophysical Journal Letters*, 806 (2015) L37 (7pp.) Published by IOP Publishing for the American Astronomical Society.

Acknowledgements

I would like to express my deepest gratitude to my supervisor, Yasushi Suto, for his continuous encouragement, inspiring discussions, and many insightful comments on my works. I feel very fortunate that I had him as my supervisor, who gave me many opportunities to broaden the possibilities for myself, always encouraged me to pursue any topic I fancy, and sincerely listened to my ideas still in their very vague forms. Above all else, I learned from his open-mindedness and his attitude to think on his own to understand something.

Chapter 4 of this thesis is a collaboration work with Othman Benomar and Hiromoto Shibahashi. Not only were they good teachers of asteroseismology, they were also very supportive collaborators, who gave me a marvelous opportunity to learn how the good collaboration works. Even after the completion of that work, they keep letting me know of stimulating research topics in the related fields, which I appreciate very much.

In Chap. 6, I was fortunate enough to have an opportunity to work with Hajime Kawahara and Sho Uehara. I enjoyed this work from start to finish, and wholeheartedly thank them for always stimulating conversations over a lot of coffee. Indeed, the work with them helped me to extend my research to the field I would have never imagined to work on otherwise. The experience also made me feel sure that the interaction between diverse personalities is an essential part of the research. I am also grateful to Takayuki Kotani and Shin'ya Yamada, for sharing cheerful experience of observations as a part of the FLEX (Fuchinobe Lightcurve EXploration) collaboration.

I would like to thank (ex-)members of the exoplanet group at UTAP (The University of Tokyo Theoretical Astrophysics group), Yuka Fujii, Teruyuki Hirano, Yuxin Xue, Shoya Kamiaka, and Masataka Aizawa. I feel fortunate that I was able to have them as good examples when I joined the group, and am thankful to them for sharing the seminars that motivated me to study exoplanets. I am particularly grateful to Teruyuki Hirano, who first introduced the *Kepler* data to me and taught me the basic ways of analyzing them. I owe it much to him that I was able to start my own research rather smoothly.

I would also like to thank all the other members of UTAP and RESCEU (Research Center for the Early Universe) for a wonderful environment where I had an enjoyable graduate student life. In particular, I would like to offer my special thanks to Masamune Oguri, Gen Chiaki, and Natsuki H. Hayatsu, with whom I spent a fun time over coffee, etc. I would like to express my respect for their efforts to make UTAP a better environment.

My sincere thanks also go to Joshua N. Winn and his group at the Massachusetts Institute of Technology, where a part of this thesis was completed, for their hospitality and helpful conversations. In completing this thesis, I was also helped by the members of my thesis committee, Profs. Takao Nakagawa, Masahiro Ikoma, Masahiro Takada, Synge Todo, Satoshi Yamamoto, and Noriko Y. Yamasaki, who gave me many constructive comments that remind me of what I need to bear in mind in my future research. I do appreciate their encouragement and support.

Last but not least, I would like to express my deep appreciation to my parents for supporting me to study at graduate school, and to my spouse, Misa, for her generosity and patience.

This thesis includes data collected by the *Kepler* mission. Funding for the *Kepler* mission is provided by the National Aeronautics and Space Administration (NASA) Science Mission Directorate. I am grateful to the entire *Kepler* team for making the revolutionary data available. Some of the data presented in this thesis were obtained from the Mikulski Archive for Space Telescopes (MAST). STScI is operated by the Association of Universities for Research in Astronomy, Inc., under NASA contract NAS5-26555. Support for MAST for non-HST data is provided by the NASA Office of Space Science via grant NNX09AF08G and by other grants and contracts. We are also grateful to Simon Albrecht and Joshua N. Winn for providing us with the radial velocity data of Kepler-25 analyzed in Chap. 4.

This thesis has made use of the NASA Exoplanet Archive, which is operated by the California Institute of Technology, under contract with NASA under the Exoplanet Exploration Program. This thesis has made use of NASA's Astrophysics Data System Bibliographic Services as well. The data analysis was in part carried out on the common use data analysis computer system at the Astronomy Data Center, ADC, of the National Astronomical Observatory of Japan.

I gratefully acknowledge the support by Japan Society for the Promotion of Science Research Fellowships for Young Scientists (No. 26-7182) and by the Leading Graduate Course for Frontiers of Mathematical Sciences and Physics.

Contents

1 **Diversity of the Extrasolar Worlds** 1
 1.1 The Overall Occurrence 1
   1.1.1 Hot Jupiters 2
   1.1.2 Super-Earths and Mini-Neptunes 3
   1.1.3 Cold and Warm Jupiters 5
   1.1.4 Eccentric Planets 5
 1.2 Planet Hunting in a Nutshell 7
   1.2.1 Direct Imaging 8
   1.2.2 Radial Velocity 8
   1.2.3 Transit 9
   1.2.4 Microlensing 12
   1.2.5 Timing 13
   1.2.6 Other Methods 14
 1.3 Directions of Stellar Spin and Planetary Orbits 15
   1.3.1 Is the Obliquity Distribution a Simple Function? 16
 1.4 Plan of This Thesis 16
 References 18

2 **Measurements of Stellar Obliquities** 21
 2.1 Definition and Terminology 21
 2.2 Obliquity from Spectroscopic Transit 22
   2.2.1 The Rossiter-McLaughlin Effect 22
   2.2.2 Doppler Tomography 25
 2.3 Obliquity from High-Precision Photometry 25
   2.3.1 Asteroseismology 25
   2.3.2 Gravity Darkening 27
   2.3.3 Spectroscopic $v \sin i_\star$ and Stellar Rotation Period ... 28
   2.3.4 Spot Anomaly 29
   2.3.5 Spot-Modulation Amplitude 29

2.4	Correlations with the System Properties		30
	2.4.1 Hot Stars (with Hot Jupiters) Have High Obliquities		30
	2.4.2 Planetary Mass Cut Off for Retrograde Planets		31
	2.4.3 Single- Versus Multi-transiting Systems		32
References			33

3 Origin of the Misaligned Hot Jupiters: Nature or Nurture? 35
3.1	High-Eccentricity Migration		36
	3.1.1 The Scenario		36
	3.1.2 Relevant Observational Issues		39
3.2	Tidal Origin of the Obliquity Trend		41
	3.2.1 Possible Evidence Against the Tidal Origin		42
3.3	Star–Disk Misalignment		44
	3.3.1 Possible Origins of Primordial Misalignment		44
	3.3.2 Obliquity Trends in the Primordial Misalignment Scenario		46
3.4	Summary and Outstanding Questions		48
	3.4.1 Are Hot Jupiters Special?		48
	3.4.2 Are All Planetary Systems Flat?		51
	3.4.3 Initial Distribution of the Star–Disk Misalignment		51
	3.4.4 Efficiency of Tides		52
References			53

4 Three-Dimensional Stellar Obliquities of HAT-P-7 and Kepler-25 from Joint Analysis of Asteroseismology, Transit Light Curve, and the Rossiter–McLaughlin Effect 55
4.1	Introduction		55
	4.1.1 A Historical View on Measurements of λ		55
	4.1.2 Aim: Determination of ψ		56
	4.1.3 Plan of This Chapter		58
4.2	Previous Measurements of Stellar Obliquities		58
	4.2.1 HAT-P-7		58
	4.2.2 Kepler-25		59
4.3	Information from Asteroseismology Analysis		59
	4.3.1 Mode Identification and Frequency Measurements		59
	4.3.2 Derivation of Fundamental Stellar Properties		61
	4.3.3 Geometry from the Rotational Splitting		64
	4.3.4 Comments on the Results for Each System		64
4.4	Joint Analysis of the HAT-P-7 System		66
	4.4.1 Analysis of Transit and Occultation Light Curves		66
	4.4.2 Joint Analysis		69
4.5	Joint Analysis of the Kepler-25 System		73
	4.5.1 Method		73
	4.5.2 Results		74

	4.6	Summary and Discussion	75
		4.6.1 HAT-P-7	75
		4.6.2 Kepler-25	77
		4.6.3 Note on the Result for Kepler-25	78
	4.7	Conclusion	78
	References		79
5	**Spin–Orbit Misalignments of Kepler-13Ab and HAT-P-7b from Gravity-Darkened Transit Light Curves**		**81**
	5.1	Introduction	81
	5.2	Method	83
		5.2.1 Model	83
		5.2.2 Data Processing	85
		5.2.3 Fitting Procedure	86
	5.3	Transit Analysis of Kepler-13Ab	86
		5.3.1 Reproducing the Results by B11	87
		5.3.2 Systematics Due to Stellar Parameters	87
		5.3.3 Joint Solution	90
	5.4	Spin–Orbit Precession in the Kepler-13A System	90
		5.4.1 Model Parameters from Each Transit	91
		5.4.2 Fit to the Observed Angles and Future Prediction	95
	5.5	Anomaly in the Transit Light Curve of HAT-P-7	99
		5.5.1 Robustness of the Observed Anomaly	100
		5.5.2 Results	100
	5.6	Summary	103
		5.6.1 Kepler-13A	103
		5.6.2 HAT-P-7	105
	5.7	Conclusion	105
	References		106
6	**Probing the Architecture of Hierarchical Multi-Body Systems: Photometric Characterization of the Triply-Eclipsing Triple-Star System KIC 6543674**		**109**
	6.1	Introduction	109
	6.2	Constraints from ETVs and Phase Curve of the Inner Binary	111
		6.2.1 ETV Analysis	111
		6.2.2 Phase-Curve Analysis	115
	6.3	Geometry and Absolute Dimensions from the Tertiary Event	117
		6.3.1 Mutual Inclination	117
		6.3.2 Relative Dimensions	118
		6.3.3 Absolute Dimensions	119
	6.4	Summary and Discussion	121
	References		122

7 Summary and Future Prospects ... 123
7.1 Summary ... 123
7.2 Future Prospects—Beyond Hot Jupiters ... 125
7.2.1 Obliquity of Longer-Period Planets Around Hot Stars ... 125
7.2.2 Warm Jupiters as Failed Hot Jupiters? ... 127
7.2.3 Stellar Obliquity Trend as the Difference in the Planetary System Architecture ... 129
References ... 130

Appendix A: Planetary Orbit ... 133

Appendix B: Summary of the Transit Method ... 139

Appendix C: Joint Posterior Distributions for the Model Parameters ... 151

Chapter 1
Diversity of the Extrasolar Worlds

Abstract The solar system consists of three different types of planets located in three distinctly separated areas. Their orbits are mostly circular and confined in a plane perpendicular to the sun's rotation axis. These regular features have led to a standard scenario for the solar system formation through the collisional growth of small rocky and icy particles (*planetesimals*) and subsequent gas accretion within a rotating circumstellar disk of gas and dust (*protoplanetary disk*). Since then, a huge diversity of exoplanets, planets orbiting stars other than the sun, has been discovered. With the steady improvements in the observational technique and the advent of new tools, we are beginning to obtain detailed information on the architectures and physical properties of those distant new worlds. Such efforts have consequently revealed that the properties of our solar system may not be the norm, and called into question what we thought we knew about the solar system. One of the goals of exoplanetary science is to understand the diversity in orbital and physical properties in a comprehensive manner. More specifically, we wish to distinguish the features of planetary systems that necessarily result from the law of Nature, from those that are sculpted by accidents specific to each system. This thesis is to deal with one aspect of those "nature and nurture" problems in exoplanetary science, which will be described in the first three chapters.

Keywords Exoplanet populations · Detection methods · Orbital architecture

1.1 The Overall Occurrence

As of April 2016, about 2000 exoplanets have been confirmed around 1200 stars. Their masses and orbital semi-major axes are shown in Fig. 1.1 by filled circles. Their colors correspond to various methods used to identify each planet, which are summarized in Sect. 1.2. The planets in our solar system are also plotted by filled diamonds for comparison. Most of the currently known exoplanets are very different from the planets in the solar system, simply because planets with similar masses and orbits to them are difficult to detect with the current technique. Even taking into

© Springer Nature Singapore Pte Ltd. 2018
K. Masuda, *Exploring the Architecture of Transiting Exoplanetary Systems with High-Precision Photometry*, Springer Theses,
https://doi.org/10.1007/978-981-10-8453-9_1

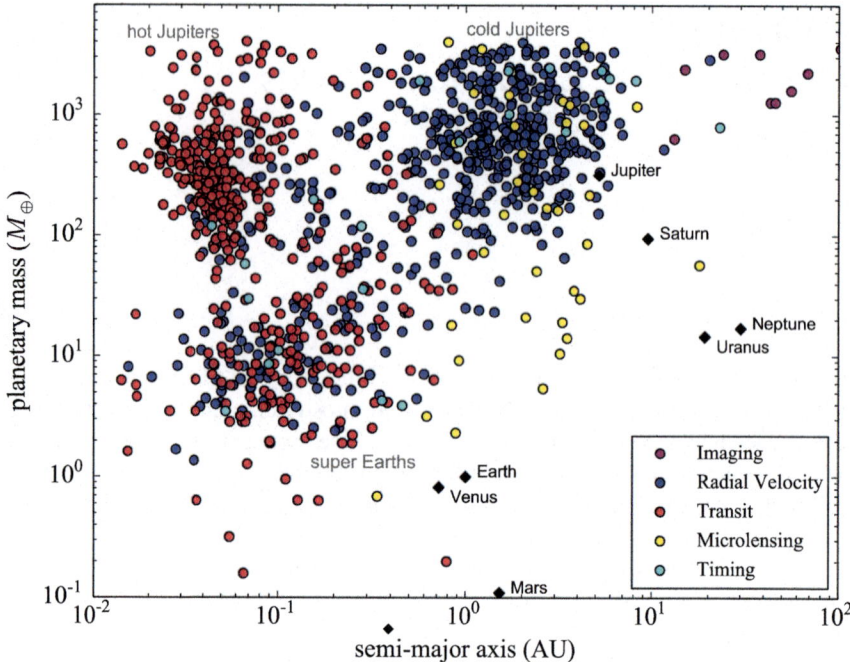

Fig. 1.1 Masses (0.1 $M_\oplus$–13$M_{\rm Jup}$) and orbital semi-major axes (0.01–100 AU) of known exoplanets as of April 14, 2016. The planets in the solar system are also shown with filled diamonds. The exoplanet data are from the confirmed planet catalog at NASA Exoplanet Archive http://exoplanetarchive.ipac.caltech.edu/index.html. The color of each circle shows a detection method by which the planet was first identified. The plotted value of the planetary mass is the "minimum" mass $M_{\rm p} \sin i$ when the true mass is not available; this is usually the case for non-transiting planets characterized with radial velocities (see Sect. 1.2)

account this detection bias, however, Fig. 1.1 already exhibits great diversity in the exoplanet property.

Three distinct populations of exoplanets show up in Fig. 1.1, which are labeled as "hot Jupiters," "cold Jupiters," and "super Earths." Below we discuss the occurrence and property of each population. We also comment on the eccentricities of exoplanets, which also show far greater diversity than the near-circular orbits in the solar system.

1.1.1 Hot Jupiters

Hot Jupiters usually refer to Jupiter-sized planets with orbital periods less than 10 days, although the thresholds are not clearly defined. They are easiest to discover with major "indirect" detection methods (i.e., radial velocity and transit; see Sect. 1.2), and the first exoplanet discovered around a Sun-like star, 51 Pegasi b, also falls into this category (Mayor and Queloz 1995). Radial velocity (RV) surveys show that they exist around roughly 1% of FGK stars (Wright et al. 2012), and seem to favor metal-rich hosts (e.g., Fischer and Valenti 2005). On the other hand, the transit survey by *Kepler* (Sect. 1.2.3) found the occurrence rate roughly half of the estimate from RV surveys (e.g., Howard et al. 2012). The origin of this possible tension is currently unclear.

In the standard scenario, formation of a giant planet involves accumulation of small particles of rock and ice into a core of $\sim 10 M_\oplus$, and subsequent accretion of surrounding H/He gas that grows the core into a gaseous giant planet (e.g., Armitage 2010). Giant planet formation via this "core accretion" process is deemed unlikely at the current location of hot Jupiters (~ 0.05 AU), where high irradiation from the host star makes the protoplanetary disk too hot for enough amount of solid particles to exist (e.g., Bodenheimer et al. 2000). The current paradigm of hot Jupiter formation, therefore, assumes that they are formed far from the host star (beyond a few AU, like Jupiter in the solar system), rather than in situ, and then "migrated" inward to their current orbits.[1]

The mechanism for the migration has been an issue of intense discussions since the discovery of the exoplanet, and is still under debate. Lin et al. (1996) proposed a mechanism usually referred to as the "disk migration," in which a giant planet opens a gap in its natal protoplanetary disk and is transported inward with the viscous diffusion of the disk over the timescale of $\sim 10^5$ yr (see, e.g., Lubow and Ida 2011, for details). This scenario, however, is unlikely to explain some of the known properties of hot Jupiters, such as the high eccentricity and spin–orbit misalignment, as will be discussed later in this chapter. The fact motivated another channel of migration including the violent few-body dynamical processes, sometimes referred to as "high-eccentricity migration." The details of this scenario will be discussed in Sect. 3.1 in Chap. 3.

1.1.2 Super-Earths and Mini-Neptunes

In the solar system, no planets between the sizes of Earth and Uranus ($15 M_\oplus$, $4 R_\oplus$) exist. In exoplanetary systems, on the other hand, many have been found in this mass and radius range (and slightly above, up to $\sim 30 M_\oplus$), as illustrated in Figs. 1.1 and 1.2. They are called "super-Earths" or "mini-Neptunes,"[2] and actually the most abundant among the known exoplanet populations, despite that they do not exist in the solar system.[3] The transit survey by the *Kepler* space telescope (Sect. 1.2.3)

---

[1] Note, however, that the possibility of in-situ formation is recently revisited (Boley et al. 2016; Batygin et al. 2016), motivated by the discovery of many super Earths (Sect. 1.1.2) on close-in orbits, which, if formed before the dispersal of the gas disk, could potentially grow into hot Jupiters.

[2] The two names are often used rather loosely without referring to their physical properties, as their internal structures are usually not very well constrained.

[3] At least they are not known, or confirmed, to exist; the predicted mass of "Planet Nine," a hypothetical planet in the outer solar system, may be in this range (Batygin and Brown 2016).

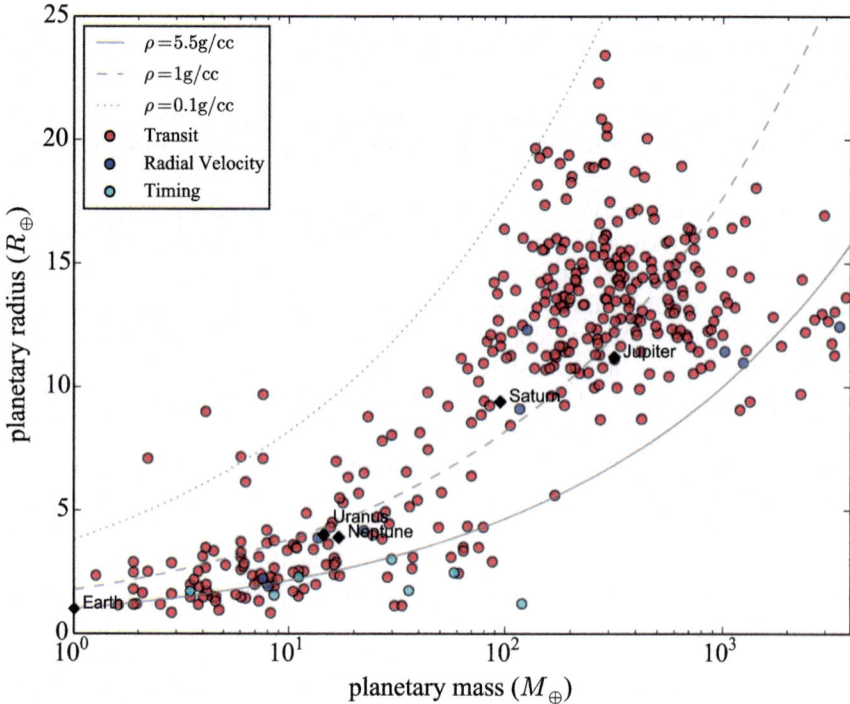

Fig. 1.2 Masses ($1\,M_\oplus$–$13\,M_{\text{Jup}}$) and radii ($<2.2\,R_{\text{Jup}}$) of known exoplanets as of May 2, 2016. The planets in the solar system are also shown with filled diamonds. The exoplanet data are from the confirmed planet catalog at NASA Exoplanet Archive http://exoplanetarchive.ipac.caltech.edu/index.html. The color of each circle shows a detection method by which the planet was first identified. Note that planets with known radii are all transiting, although some of them may be first identified with radial velocities (several blue circles in the plot). Thus the masses plotted in this figure are basically the true masses without the ambiguity due to the unknown orbital inclination

has revealed that planets with $1\,R_\oplus < R_p < 4\,R_\oplus$ and periods $\lesssim 100$ days exist around ∼50% of Sun-like stars (e.g., Fressin et al. 2013), and are often found in compact multi-planetary systems, where multiple planets reside in closely-packed orbits (Lissauer et al. 2011b; Fabrycky et al. 2014).

Most of the currently known super-Earths have orbits smaller than that of Venus. While this may indicate that super-Earths, like hot Jupiters, experienced smooth inward migration through the interaction with the gas disk (e.g., Goldreich and Tremaine 1980), in situ formation from a more massive disk than assumed in the solar-system model (Hayashi 1981) is also discussed as a viable possibility (e.g., Raymond and Cossou 2014), given that they have relatively smaller masses and that they do not need to be formed before the disk dispersal, unlike Jupiter-sized planets.

The planets in this mass/radius range are known to exhibit a wide range of physical properties, with their mean densities spanning over almost two orders of magnitudes, from less than $0.1\,\text{g}\,\text{cm}^{-3}$ (Masuda 2014) to more than the value ($5.5\,\text{g}\,\text{cm}^{-3}$) of Earth

(Carter et al. 2012). Some of them seem to be scaled-up version of Earth consisting of an iron core overlaid with a silicate mantle (e.g., Weiss et al. 2016), while others need significant fractions of gas envelopes to account for the observed mass and radius (e.g., Lissauer et al. 2013). Both the statistical analysis of observational data and theoretical modeling of interior structures suggest that the dividing line between the rocky planets and planets with gaseous envelopes (i.e., "physical" distinction between super-Earths and mini-Neptunes) exists around $1.6\,R_\oplus$ (Weiss and Marcy 2014; Lopez and Fortney 2014; Rogers 2015).

1.1.3 Cold and Warm Jupiters

Given the current precision of the radial velocity measurements ($\lesssim 1\,\mathrm{m\,s^{-1}}$), Jupiter-mass planets around several AU from the host star (i.e., Jupiter analogues) are readily detectable (see Eq. 1.4), if monitored for a sufficiently long ($\lesssim 10\,\mathrm{yr}$) duration (e.g., Vogt et al. 2014). The Doppler surveys performed in the past decade showed that the occurrence rate of such "cold Jupiters" is roughly 10% for FGK stars (see, e.g., Cumming et al. 2008, who found the occurrence rate of 10.5% for $P < 5.5\,\mathrm{yr}$ and $M_\mathrm{p} = 0.3\text{--}10\,M_\mathrm{Jup}$ from eight-year measurements).

The same surveys (Udry et al. 2003; Cumming et al. 2008) have also reported the lack of Jupiter-mass planets with intermediate orbital radii (often called "warm Jupiters"), below the rise of occurrence around 1 AU. Indeed, this so-called "period valley" seems to make a natural distinction between hot and cold Jupiters in Fig. 1.1. The origin of the period valley and warm Jupiters is not understood and still debated actively (e.g., Dawson and Murray-Clay 2013; Dong et al. 2014; Dawson and Chiang 2014; Huang et al. 2016).

1.1.4 Eccentric Planets

Another notable feature of exoplanets, not captured in Fig. 1.1, is their eccentricity distribution. While the orbits of solar-system planets are mostly circular, except for Mercury with $e = 0.21$, exoplanets exhibit a far wider range of orbital eccentricities, which are plotted against their semi-major axes in Fig. 1.3. For example, HD 80606b, the planet with one of the largest eccentricities, resides in an almost parabolic orbit with $e = 0.93$ (Naef et al. 2001). The possible origins for such high eccentricities are the gravitational scattering between multiple planets (e.g., Lin and Ida 1997; Chatterjee et al. 2008; Jurić and Tremaine 2008) and long-term gravitational perturbation from a companion star (e.g.,Takeda and Rasio 2005); these processes will be further discussed in Sect. 3.1.

Figure 1.3 shows that the maximum eccentricity increases with increasing semi-major axis (e.g., Butler et al. 2006), with its upper envelope consistent with a constant value of the orbital pericenter distance $q = a(1-e)$. This suggests that tidal dissipa-

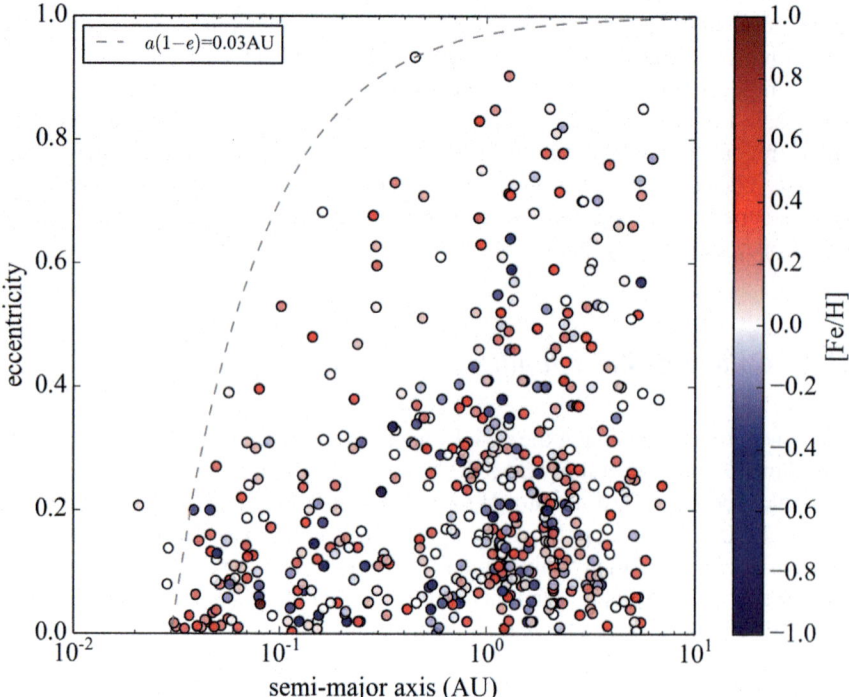

Fig. 1.3 Orbital eccentricity versus semi-major axis of known exoplanets detected with radial velocities. The exoplanet data are from the confirmed planet catalog at NASA Exoplanet Archive http://exoplanetarchive.ipac.caltech.edu/index.html. The dashed line corresponds to the pericenter distance of 0.03 AU. The color of each circle corresponds the metallicity of the planet's host star

tion plays a role; variation in the star–planet distance over an eccentric orbit causes the time-dependent distortions of the two bodies, which are eventually dissipated within them. The resulting loss of orbital energy leads to the orbital circularization, whose timescale depends sensitively on the minimum star–planet distance q (e.g., Correia and Laskar 2010). In fact, the same mechanism may also be responsible for the formation of hot Jupiters, as will be detailed in Sect. 3.1.

Figure 1.4 illustrates another feature that larger planets are more likely to have larger eccentricities (e.g., Wright et al. 2009). Wright et al. (2009) also found that planets in multi-planetary systems tend to have smaller eccentricities; this trend may be associated with the correlation with planetary size, as the smaller planets are more often found in multi-planetary systems. In addition, Dawson and Murray-Clay (2013) identified that giant planets with semi-major axes 0.1 AU–1 AU, i.e., warm Jupiters, around metal-rich stars with [Fe/H] > 0 have higher eccentricities than those around metal-poor stars with [Fe/H] < 0; the trend can be seen in Fig. 1.3 as the lack of blue circles in the corresponding regime. This trend, along with the

1.1 The Overall Occurrence

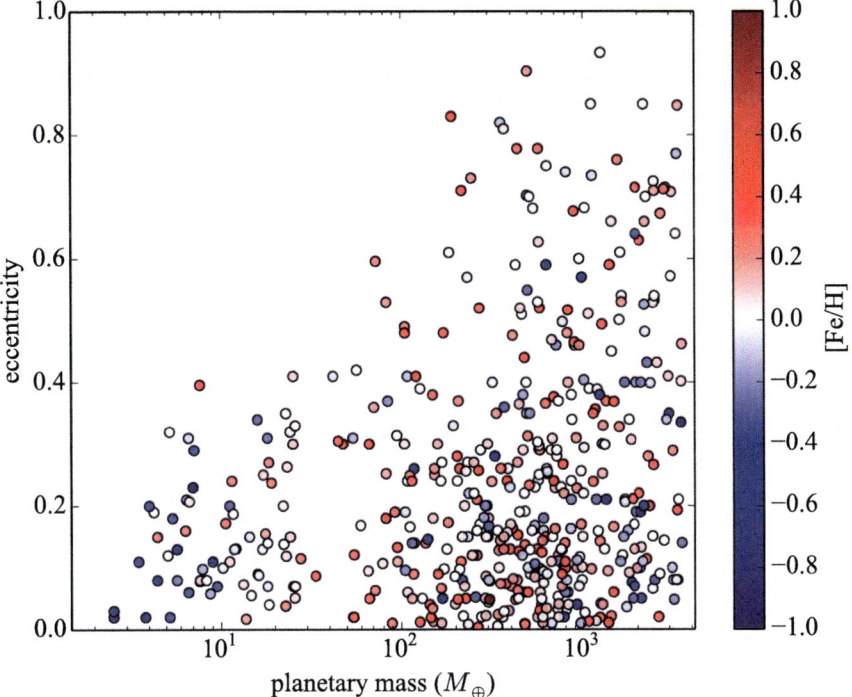

Fig. 1.4 Orbital eccentricity versus mass of known exoplanets detected with radial velocities. The exoplanet data are from the confirmed planet catalog at NASA Exoplanet Archive http://exoplanetarchive.ipac.caltech.edu/index.html. The mass is basically $M_p \sin i$ but M_p is plotted if available. The color of each circle corresponds to the metallicity of the planet's host star

above eccentricity–mass correlation, may be the sign of eccentricity excitation due to planet–planet scattering, as more giant planets could form around more metal-rich stars.

1.2 Planet Hunting in a Nutshell

In the following, we briefly comment on each of the methods used to detect exoplanets in Fig. 1.1. The "direct imaging" method is to capture the light from an exoplanet directly, while the others infer the existence of planets by observing the radiation from their host stars and hence are called indirect methods.

1.2.1 Direct Imaging

The most direct way to detect exoplanets around other stars is to actually *image* them, as can be done for the planets in the solar system. This requires an extraordinary effort to overcome the contrast between a planet and its host star.

Suppose, for example, that we try to detect the light reflected by a "second Earth" around another star, with radius R_p and semi-major axis a. The flux ratio f between the planet and its host star is then given by

$$f_{\text{reflection}} = A_b \frac{\pi R_p^2}{4\pi a^2} = 10^{-10} \left(\frac{A_b}{0.3}\right) \left(\frac{R_p}{R_\oplus}\right)^2 \left(\frac{a}{1\,\text{AU}}\right)^{-2}, \quad (1.1)$$

where $A_b = 0.3$ is the bond albedo of the Earth. The ratio corresponds to the magnitude difference of about 25. To detect such a faint source even as an isolated object, deep exposures are required.

A more promising approach is to observe the thermal emission from an exoplanet at longer wavelengths. Assuming that both the planet and star are blackbody radiators, the contrast in the thermal flux is given by

$$f_{\text{thermal}} = \left(\frac{R_p}{R_\star}\right)^2 \frac{\exp(hc/k_B T_p \lambda) - 1}{\exp(hc/k_B T_\star \lambda) - 1}, \quad (1.2)$$

where k_B is the Boltzmann constant, h is the Planck constant, and c is the speed of light in vacuum. For $\lambda = 10\,\mu$m, at which the planet at $a = 1$ AU around a Sun-like star is brightest, the contrast amounts to $f_{\text{thermal}} \sim 10^{-6} \gg f_{\text{reflection}}$ for $R_p = R_\oplus$ and $R_\star = R_\odot$. Even in the infrared, the problem of resolution needs to be overcome; to resolve $a = 10$ AU at the distance of 10 pc, a telescope with a diameter of ~ 5 m is required.

For these reasons, the direct imaging method has mainly discovered young, self-luminous planets far from the host star (Fig. 1.1). The mass of each planet is estimated from the system age and planet luminosity via the cooling model of young Jupiters and brown dwarfs (e.g., Baraffe et al. 2003). Basically, more massive planets are brighter at a given age because they cool down more slowly. The planetary mass thus estimated is usually more reliable for older systems because the model becomes largely independent from the unknown initial condition (Kuzuhara et al. 2013).

1.2.2 Radial Velocity

The acceleration of the host star induced by its planet's gravity can be detected by measuring the stellar radial velocity (RV) with spectroscopy. The RV variation of a planet-hosting star, with respect to the barycentric motion of the system, is given by

1.2 Planet Hunting in a Nutshell

$$v_\star(t) = K_\star \left(\cos[\omega + f(t)] + e\cos\omega\right), \quad (1.3)$$

where e, ω, and f are the eccentricity, argument of pericenter, and true anomaly of the planetary orbit relative to the star (cf. Appendix A), and the RV semi-amplitude $K_\star$ is given by

$$\begin{aligned}K_\star &= \left(\frac{M_{\rm p}}{M_\star + M_{\rm p}}\right)\frac{na\sin i}{\sqrt{1-e^2}} = \left(\frac{2\pi GM_\star}{P}\right)^{1/3}\frac{M_{\rm p}/M_\star}{(1+M_{\rm p}/M_\star)^{2/3}}\frac{\sin i}{\sqrt{1-e^2}} \\ &= \frac{28\,{\rm m\,s^{-1}}}{\sqrt{1-e^2}}\frac{M_{\rm p}\sin i}{M_{\rm Jup}}\left(\frac{M_\star + M_{\rm p}}{M_\odot}\right)^{-2/3}\left(\frac{P}{1\,{\rm yr}}\right)^{-1/3},\end{aligned} \quad (1.4)$$

with i being the orbital inclination relative to the sky plane (cf. Appendix A).

The RV time series sufficiently sampled over the whole phase yields e, ω, orbital phase, orbital period, and $K_\star$. The RV semi-amplitude, combined with P and e, translates into the constraint on $M_{\rm p}\sin i/(M_\star + M_{\rm p})^{2/3}$. Given the stellar mass, therefore, one obtains a dynamical constraint on the planetary mass through the combination $M_{\rm p}\sin i$. The quantity is usually called "minimum" planetary mass, since the true planetary mass $M_{\rm p}$ is always larger than $M_{\rm p}\sin i$. Because the RV method alone is not sensitive to the orbit direction at all, it is only possible for some special cases (e.g., transiting planets with $i \simeq \pi/2$) that the true planetary mass is obtained from RVs. The masses of RV planets in Fig. 1.1 are either of this minimum mass or the true mass if available.[4]

Equation (1.4) shows that the signal scales as $M_{\rm p}P^{-1/3}$ or $M_{\rm p}a^{-1/2}$, and so the technique is biased toward more massive and shorter-period planets (see Fig. 1.1). While it is also true that $K_\star$ is larger for more eccentric orbits, the sensitivity dependence on eccentricity is more complex because the finer sampling around the pericenter, where the RV (only) exhibits significant variations for a highly eccentric orbit, is required to detect such a planet (Cumming 2004).

1.2.3 Transit

A planet, if viewed edge-on, periodically passes in front of the stellar disk to produce periodic "dips" in the stellar flux. Such an eclipse by a planet, usually called "planetary transit," provides a way to detect exoplanets with the photometric observation. Below we summarize its basic concepts and leave the more detailed discussion of the method in Appendix B.

---

[4] Strictly speaking, it is always possible that the object with $M_{\rm p}\sin i$ comparable to that of Jupiter is actually a substellar object (e.g., Sahlmann et al. 2011). It should be noted, however, that the minimum mass is a priori close to the true mass if the orbit direction is isotropic; for example, the probability that the true mass is larger than the twice of the minimum mass is only 13%, and the true mass is larger than the minimum mass only by a factor of $4/\pi$ on average.

A great advantage of the transit is that it reveals the planetary radius, which is inaccessible with any other detection methods. Since the planet can be regarded as a dark disk in the usual photometry, the depth of the dip, or the transit depth δ, is essentially given by the ratio of the area of the planetary disk to that of the star:

$$\delta = \left(\frac{R_p}{R_\star}\right)^2. \tag{1.5}$$

The transit observation also yields an extremely precise orbital period, whose precision is improved linearly with the time baseline. As mentioned above, the RV follow-up of transiting planets also leads to the true planetary mass, which is otherwise difficult to obtain. In fact, the process is usually essential to confirm a transiting object to be a genuine planet, because the transit light curve does not reveal the mass of the object (but see below for exceptions). Transits also provide the precious opportunity to study exoplanet atmosphere by analyzing the light that grazes, or is emitted/reflected by, the upper atmosphere of the planet and reaches to us (transmission/occultation spectroscopy).

As such, transiting planets are advantageous targets for detailed characterization in many aspects. The problem is how to find them, because the transit can be observed only for a short duration, and for systems with suitable geometry. For a randomly oriented planetary orbit, the probability that a given planet transits as seen from an observer on Earth is roughly $R_\star/a$; this is $\sim 10\%$ for the most close-in planets, and only 0.5% for an Earth-like planet around a Sun-like star. Moreover, even if the system does have the geometry to exhibit transits, we need to observe the star at the right time, since the transit lasts for a fraction $R_\star/\pi a$ of the whole orbital period; typical duration of the transit is given by[5]

$$T_0 = \frac{R_\star P}{\pi a} = 13\,\text{hr} \left(\frac{P}{1\,\text{yr}}\right)^{1/3} \left(\frac{\rho}{\rho_\odot}\right)^{-1/3} = 4\,\text{hr} \left(\frac{P}{10\,\text{days}}\right)^{1/3} \left(\frac{\rho}{\rho_\odot}\right)^{-1/3}. \tag{1.6}$$

Combining all these together, the transit method is strongly biased toward short-period planets, even more than the RV method. This drawback is overcome by the continuous monitoring of a large number of stars from space, as has been made possible by *CoRoT* (Baglin et al. 2006) and *Kepler* (Borucki et al. 2010) space telescopes, which have brought ample photometric data with unprecedented precision and time coverage.

This thesis is largely based on the data of transiting systems collected by *Kepler*. Below we give a brief overview of the telescope and another detection and characterization method made possible by *Kepler*.

---

[5]We use Kepler's third law divided by the stellar radii cubed, $4\pi^2(a/R_\star)^3/P^2 = GM_\star/R_\star^3$, to derive this scaling. The relation shows that the timescale of the transit essentially fixes the density of the system, which is the only physical dimension constrained from the light curve alone (see also Appendix B).

1.2.3.1 The *Kepler* Space Telescope

The *Kepler* space telescope is implemented with a differential photometer with a 115 deg^2 field of view and continuously monitored the brightness of $\sim$160000 stars located in the constellations of Cygnus and Lyra (Borucki et al. 2010). During the main mission over four years between 2009–2013, *Kepler* has discovered more than 4000 transiting planet candidates, whose orbital periods and radii are estimated from the transit light curve. Among these candidates, more than 1000 are "confirmed" to be genuine planets as of April 2016, by determining their masses and/or showing that they are highly unlikely to be astrophysical false positives: phenomena that mimic the transit-like signal (mostly blended eclipsing binaries).

The preliminary stellar parameters of the target stars, including magnitudes in different bands, effective temperature, surface gravity, and metallicity, are listed in the *Kepler* Input Catalog (KIC, Brown et al. 2011) available at the MAST archive,[6] and each star is given a KIC number. If the periodic dips are found in the light curve of a star, the star is listed up as a *Kepler* Object of Interest (KOI, Coughlin et al. 2015), and given a KOI number of the form "KOI-xxxxx". The source object of the transit-like signal is also given a KOI name of the form "KOI-xxxxx.yy", where yy distinguishes multiple signals found for one star, and is assigned beginning from 01 in the order of detection. Finally, if the transit-like signal is confirmed to be due to a genuine planet, the system is given a Kepler number like "Kepler-zzz," and each planet is assigned a lower-case letter beginning from "b," basically in the order of increasing distance from the host star.[7]

Kepler observes a target in two cadences, long cadence (LC; 29.4 min) and short cadence (SC; 58.35 s). The LC photometry has been obtained for all the targets, while the SC data exist for a selected set of targets (mainly KOIs, and sometimes include eclipsing binaries). The photometric precision integrated over 6.5 hr (comparable to the typical transit duration; Eq. 1.6) is estimated to be better than 100 ppm ($= 10^{-4}$) for a $V \lesssim 14$ star typical in the *Kepler* target (Van Cleve et al. 2016). The value is comparable to the transit depth expected for a Sun-Earth system with $R_p/R_\star \simeq 0.01$.

The primary mission ended in the summer of 2013 due to the loss of two reaction wheels. Adopting a tricky way of operating the spacecraft and maintaining its pointing, however, the telescope has now been reused for the *K2* mission to sequentially observe different patches of the sky along the ecliptic for shorter ($\sim$80 days) durations (Howell et al. 2014) and is still in operation as of May 2016.

---

[6] https://archive.stsci.edu/index.html.

[7] The letter "a" is reserved for the central star. If the host star forms a multi-stellar system, the capital letter follows after the Kepler number to specify the planet-hosting component (like "Kepler-16A b"). The order of planet letters is sometimes irregular because inner planets may be found after the outer one(s) in some cases.

1.2.3.2 Transit Timing Variation

The long-term, continuous monitoring by *Kepler* has made it possible to detect gravitational interaction between multiple planets in the same system. The temporal modulation of the orbital period of a transiting planet, measured very precisely from the interval between the successive transits, allows for the detection of another non-transiting planet and/or detailed characterization of the transiting ones. The method is referred to as the transit timing variation (TTV, Miralda-Escudé 2002; Holman and Murray 2005; Agol et al. 2005).

TTVs allow for the confirmation of transiting planet candidates identified by *Kepler*, without (often demanding) RV observations. Indeed, the technique is often the *only* option for many of the multi-transiting systems that are too faint to observe RVs with a sufficient precision in a reasonable amount of time, and hence essential for maximizing the scientific yield from the *Kepler* data. Moreover, TTVs serve as a valuable probe of the diversity of low-mass planets (see Sect. 1.1.2) owing to its sensitivity down to Earth mass or even smaller (Jontof-Hutter et al. 2015), and have discovered planetary systems with unique properties (e.g., Lissauer et al. 2011a; Sanchis-Ojeda et al. 2012; Carter et al. 2012; Masuda 2014). An attempt has also been made to use such precise timing data for many planetary systems to constrain the time-variation of a fundamental constant (Masuda and Suto 2016).

1.2.4 Microlensing

When a foreground star happens to pass very close to our line of sight to a more distant background star, the foreground star acts as a lens to split the background star into several images. The split images are typically unresolved and simply observed as a temporal magnification of the background star (*microlensing*) because surface brightness of the source is unaffected by lensing.[8] If the lens object hosts a planet, it causes an additional, short-lived magnification feature, which can be used to infer the planet-to-lens mass ratio and sky-projected lens–planet distance (see, e.g., Gaudi 2010 for are view).

To significantly perturb the image, planets need to be located close to the Einstein ring, whose angular radius is given by

$$\theta_E = \sqrt{\frac{4GM_L}{c^2}\left(\frac{1}{d_L} - \frac{1}{d_S}\right)}, \quad (1.7)$$

where M_L is the lens mass, and d_L and d_S are the distances to the lens and source, respectively. The configuration typical for planet detection is that the source lies in the Galactic bulge (with $d_S = 8$ kpc), while the lens is in the Galactic disk ($d_L \sim 4$ kpc).

---

[8] See, e.g., Sect. 9.2 of Weinberg (2008).

1.2 Planet Hunting in a Nutshell

The corresponding angular and physical Einstein radii are

$$\theta_E = 550\,\mu\text{arcsec} \times \left(\frac{M_L}{0.3 M_\odot}\right)^{1/2} \left(\frac{d_S}{8\,\text{kpc}}\right)^{-1/2} \left(\frac{d_S/d_L}{2} - 1\right)^{1/2} \quad (1.8)$$

and

$$\theta_E d_L = 2.2\,\text{AU} \times \left(\frac{M_L}{0.3 M_\odot}\right)^{1/2} \left(\frac{d_S}{8\,\text{kpc}}\right)^{1/2} \left[\frac{d_L/d_S - (d_L/d_S)^2}{0.25}\right]^{1/2}. \quad (1.9)$$

The lensed multiple images are therefore typically unresolved, as mentioned above, and the microlensing method is typically sensitive to planets around the snow line. It can also be shown that the duration and amplitude of the magnification of planetary origin only weakly depends on the planetary mass (Gaudi 2010). For these reasons, the microlensing provides a unique probe of the parameter region that is out of reach of other methods; this is clearly illustrated by yellow circles in Fig. 1.1. On the other hand, it is difficult to follow-up the planet found by microlensing for further characterization. Hence this methodology is suited to discussing the statistics, rather than detailed characterization of individual systems.

1.2.5 Timing

Suppose that a planet-hosting star has a "clock" that can be read by a distant observer like us. Due to the finite speed of light, the gravitational acceleration induced by the planet causes the temporal delay and speed up of the clock, whose amplitude Δt is given by

$$\Delta t = \frac{a}{c}\frac{M_p}{M_\star + M_p} = 1.5\,\text{ms} \times \left(\frac{a}{\text{AU}}\right)\left(\frac{M_p}{M_\oplus}\right)\left(\frac{M_\star + M_p}{M_\odot}\right)^{-1}. \quad (1.10)$$

The estimate shows that planets as light as Earth can be detected if the host "star" is an astrophysical object with an ultra-precise clock, i.e., a pulsar. This was indeed the case for a planetary system around the millisecond pulsar PSR B1257+12 (Wolszczan and Frail 1992). It is worth noting that the discovery prevails the detection of the "first" exoplanet around a normal star in 1995.

If the companion is a stellar-mass object, less precise clocks are also useful. The examples of such clocks include eclipsing binaries (see Chap. 6), stars exhibiting coherent pulsations (e.g., Shibahashi and Kurtz 2012), and inner short-period transiting planets. In addition to the above "light-travel time" effect, the clock can be physically delayed due to the gravitational tidal force. It is usually the latter effect that is referred to as TTVs in the exoplanet literature, since the light-travel time effect due to a planetary-mass object is hard to detect without such an ultra-precise clock as a pulsar. In Fig. 1.1, the planets detected with these methods, including TTVs, are marked as "timing" altogether.

1.2.6 Other Methods

1.2.6.1 Astrometry

High-precision astrometry can directly observe the sky-plane stellar motion caused by a planet. The angular resolution $\Delta\theta$ required to detect a planet around a star at distance d is

$$\Delta\theta = \frac{a}{d}\frac{M_p}{M_\star} = 5 \times 10^{-4} \text{ arcsec} \left(\frac{a}{5\,\text{AU}}\right)\left(\frac{d}{10\,\text{pc}}\right)^{-1}\left(\frac{M_p}{M_\text{Jup}}\right)\left(\frac{M_\star}{M_\odot}\right)^{-1}. \quad (1.11)$$

The precision required for the exoplanet detection is currently hard to achieve, and so the applications of this technique has so far been limited to sub-stellar mass objects including brown dwarfs (e.g., Sahlmann et al. 2013). It is expected that GAIA mission by the European Space Agency will be able to detect planetary mass companions with astrometry.

Astrometric detection of the sky-plane orbit as a function of time allows for the reconstruction of the full orbit in three-dimensions. If successfully applied to planets with RV measurements, therefore, the true mass of the companion can be obtained without the ambiguity of orbital inclinations. Such analyses have been performed for some "Jupiter-mass planets" detected with RVs to reveal their non-planetary nature by showing that the orbit is close to face-on (e.g., Sahlmann et al. 2011).

1.2.6.2 Orbital Brightness Modulation

Close-in planets generate the brightness modulation of the star–planet system in phase with the orbital motion, which is detectable only with high-precision space-based photometry (e.g., Shporer et al. 2011). The modulation usually consists of the ellipsoidal variation caused by the tidal distortion of the star (Morris and Naftilan 1993), Doppler beaming due to the stellar reflex motion (Loeb and Gaudi 2003), and the phase modulation of the light emitted and/or reflected by the planet, among which the first two can be used to constrain the companion's mass. Several transiting planets were confirmed by examining this orbital brightness modulation (e.g., Faigler et al. 2013).

1.3 Directions of Stellar Spin and Planetary Orbits

Another notable feature of exoplanets, which this thesis focuses on, is the misalignment between the axes of stellar spin and planetary orbit (spin–orbit misalignment). In the solar system, the equator of the sun and orbital planes of the eight planets are aligned within 7°. This regularity has been the basis of the standard paradigm of planet formation from the rotating protoplanetary disk, which has a long history that dates back to Nebular Hypothesis by Kant and Laplace.

Observations, however, have shown that it is not always the case in exoplanetary systems, especially for hot Jupiters (Sect. 1.1.1), as illustrated in Fig. 1.5. In this figure, each filled circle corresponds to each planetary orbit, where the distance from the star is proportional to the logarithm of the orbital period or semi-major axis, and the position angle shows the points where each orbit crosses the sky plane from this side of the paper. The latter angle, denoted as λ, is equivalent to the sky projection of the *stellar obliquity*, the angle between the stellar spin and planetary orbital axes, and is measured spectroscopically via the so-called Rossiter-McLaughlin (RM) effect (see Sect. 2.2 for more detail). While we do see some clustering of planets with spin–orbit alignments comparable to the solar system, more than one third of the sample systems have significantly misaligned orbits (see also the histogram in Fig. 1.6), and some even revolve in the opposite directions to the stellar rotation (retrograde orbit).

Currently, most of such measurements are for hot Jupiters, and the misalignment is often attributed to their "violent" migration including few-body dynamical processes, as will be discussed in Chap. 3. This scenario, however, is still incomplete in many aspects, and a part of the observed misalignments may possibly be a generic feature

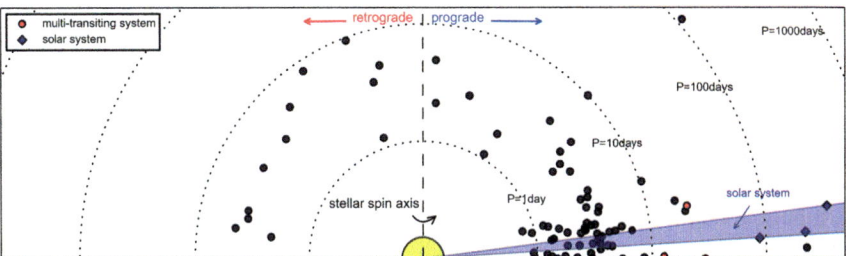

Fig. 1.5 Summary of the measurements of the sky-projected obliquity λ (Sect. 2.1) as of April 2016. The list of the systems is based on Holt-Rossiter-McLaughlin Encyclopedia by René Heller (http://www2.mps.mpg.de/homes/heller/.), and the system parameters are retrieved from NASA Exoplanet Archive (http://exoplanetarchive.ipac.caltech.edu.), although the spurious measurements of λ are omitted here in a spirit similar to Albrecht et al. (2012). Each filled circle corresponds to the descending node of each planet, where the planetary orbit crosses the sky plane from this side of the paper to the other. Distance from the central star (yellow circle at the origin) is proportional to the logarithm of the orbital period. Blue diamonds are the values for some of the planets in the solar system, with the blue-shaded region showing the entire range of the stellar obliquity for the solar-system planets. Red circles are planets in multi-transiting systems, which will be further discussed in Sect. 2.4.3

Fig. 1.6 Histogram of the λ measurements in Fig. 1.5. The vertical axis is normalized so that the integral over the entire range of λ be unity

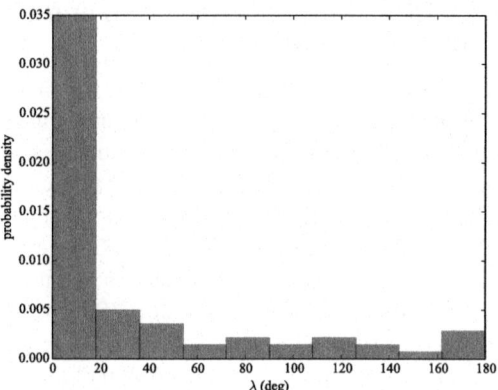

of exoplanetary system that is not related to the migration. If this is the case, the good spin–orbit alignment in the solar system may not be the norm but simply an initial condition. This thesis describes an effort to understand the origin of the spin–orbit misalignment and its relationship to the dynamical history of diverse exoplanets.

1.3.1 Is the Obliquity Distribution a Simple Function?

The first question to ask may be whether the distribution of λ could be compatible with a simple function of the true stellar obliquity ψ. For that purpose, the geometric effect needs to be taken into account because λ is the sky projection of ψ (see Fig. 2.1). Fabrycky and Winn (2009) answered this question negatively, showing that the observed distribution of λ cannot be well described with an isotropic function of ψ nor a Fisher distribution on a sphere with a constant dispersion. The failure of a single distribution is essentially because the observed λ has both a significant peak around $\lambda = 0°$ and a long tail (see Fig. 1.6), which are difficult to be reconciled simultaneously. The result hints a rather complex origin of the observed obliquity distribution; that will be the main topic of discussion in the following chapters.

1.4 Plan of This Thesis

This thesis deals with the origin of the spin–orbit misalignment from an observational viewpoint. While the measurements of stellar obliquities presented in Fig. 1.5 have mainly been performed using a spectroscopic technique (i.e., the RM effect), here we focus on the methods using high-precision photometric data obtained by the *Kepler* space telescope. We will show how such methods help us to understand the origin of the spin–orbit diversity in exoplanetary systems by providing the information com-

1.4 Plan of This Thesis

plementary to the spectroscopic one and by extending the obliquity measurements to systems that were beyond the scope of the RM effect.

We begin with summarizing the methods of obliquity measurements and current observational results in Chap. 2. Here we also examine the correlation between the stellar obliquity and other system properties, especially the one with the stellar effective temperature, which will give clues for understanding the origin of high obliquities. Chap. 3 describes two contrasting theoretical scenarios for the origin of the spin–orbit misalignment. The first one is the "high-eccentricity migration," which produces large spin–orbit misalignments, as well as large orbital eccentricities, in the course of the migration of hot Jupiters. A great deal of discussion will also be devoted to the interpretation of the trend with the effective temperature in the context of this scenario, referring to its major drawbacks as well. We then comment on the second scenario at the other extreme, that the observed spin–orbit misalignment is of primordial origin, rather than the outcome of the violent dynamical migration of hot Jupiters. The chapter is closed by listing important questions that need to be addressed for settling this nature and nurture problem, emphasizing the need of extending the obliquity measurements to planets other than hot Jupiters.

In Chaps. 4 and 5, we present measurements of stellar obliquities using the high-precision photometric data obtained by the *Kepler* spacecraft. In Chap. 4, we use asteroseismology for the measurements of the true obliquity, rather than the sky-projected obliquity usually obtained from the spectroscopic technique. We present, for the first time, a consistent procedure for the joint analysis of spectroscopic and photometric data, with the applications to two important systems. Chap. 5 focuses on another methodology of gauging the obliquity using gravity darkening exhibited by fast-rotating stars. We present an updated analysis of this phenomenon in the Kepler-13A system, where the gravity-darkening method and the spectroscopic one was known to disagree. We provide a possible solution for this discrepancy and propose a procedure to test this conclusion with future follow-up observations. We also apply the same technique to the HAT-P-7 system for the first time, and give a cross-validation of the result in Chap. 4.

In Chap. 6, photometric data are used for characterizing a triple-star system in a hierarchical configuration. Here the modeling of eclipse light curves is combined with the dynamical modeling of the multi-body gravitational interaction to yield precise masses and radii of three stars in the system essentially from the photometric data alone. The analysis presented here does not only expand the potential of high-precision photometric data, but also serves as a useful test bed for characterizing hierarchical multi-planetary systems that would serve as direct evidence for the dynamical interaction and migration scenario as discussed in Chap. 3.

Finally, Chap. 7 summarizes the results and concludes. Possible directions of future studies are also presented.

References

E. Agol, J. Steffen, R. Sari, W. Clarkson, MNRAS **359**, 567 (2005)
S. Albrecht, J.N. Winn, J.A. Johnson et al., ApJ **757**, 18 (A12) (2012)
P.J. Armitage, *Astrophysics of Planet Formation* (2010). p. 294
A. Baglin, M. Auvergne, L. Boisnard, et al., in *COSPAR Meeting, 36th COSPAR Scientific Assembly*, vol. 36 (2006)
I. Baraffe, G. Chabrier, T.S. Barman, F. Allard, P.H. Hauschildt, A&A **402**, 701 (2003)
K. Batygin, P.H. Bodenheimer, G.P. Laughlin, ApJ **829**, 114 (2016)
K. Batygin, M.E. Brown, AJ **151**, 22 (2016)
P. Bodenheimer, O. Hubickyj, J.J. Lissauer, Icarus **143**, 2 (2000)
A.C. Boley, A.P. Granados Contreras, B. Gladman, ApJ **817**, L17 (2016)
W.J. Borucki, D. Koch, G. Basri et al., Science **327**, 977 (2010)
T.M. Brown, D.W. Latham, M.E. Everett, G.A. Esquerdo, AJ **142**, 112 (2011)
R.P. Butler, J.T. Wright, G.W. Marcy et al., ApJ **646**, 505 (2006)
J.A. Carter, E. Agol, W.J. Chaplin et al., Science **337**, 556 (2012)
S. Chatterjee, E.B. Ford, S. Matsumura, F.A. Rasio, ApJ **686**, 580 (2008)
A.C.M. Correia, J. Laskar, *Tidal Evolution of Exoplanets*, ed. by S. Seager (2010), pp. 239–266
J.L. Coughlin, F., Mullally, S.E. Thompson, et al. (2015), arXiv:1512.06149
A. Cumming, MNRAS **354**, 1165 (2004)
A. Cumming, R.P. Butler, G.W. Marcy et al., PASP **120**, 531 (2008)
R.I. Dawson, E. Chiang, Science **346**, 212 (2014)
R.I. Dawson, R.A. Murray-Clay, ApJ **767**, L24 (2013)
S. Dong, B. Katz, A. Socrates, ApJ **781**, L5 (2014)
D.C. Fabrycky, J.N. Winn, ApJ **696**, 1230 (2009)
D.C. Fabrycky, J.J. Lissauer, D. Ragozzine et al., ApJ **790**, 146 (2014)
S. Faigler, L. Tal-Or, T. Mazeh, D.W. Latham, L.A. Buchhave, ApJ **771**, 26 (2013)
D.A. Fischer, J. Valenti, ApJ **622**, 1102 (2005)
F. Fressin, G. Torres, D. Charbonneau et al., ApJ **766**, 81 (2013)
B.S. Gaudi (2010), arXiv:1002.0332
P. Goldreich, S. Tremaine, ApJ **241**, 425 (1980)
C. Hayashi, Progress of theoretical physics supplement **70**, 35 (1981)
M.J. Holman, N.W. Murray, Science **307**, 1288 (2005)
A.W. Howard, G.W. Marcy, S.T. Bryson et al., ApJS **201**, 15 (2012)
S.B. Howell, C. Sobeck, M. Haas et al., PASP **126**, 398 (2014)
C. Huang, Y. Wu, A.H.M.J. Triaud, ApJ **825**, 98 (2016)
D. Jontof-Hutter, J.F. Rowe, J.J. Lissauer, D.C. Fabrycky, E.B. Ford, Nature **522**, 321 (2015)
M. Jurić, S. Tremaine, ApJ **686**, 603 (2008)
M. Kuzuhara, M. Tamura, T. Kudo et al., ApJ **774**, 11 (2013)
D.N.C. Lin, P. Bodenheimer, D.C. Richardson, Nature **380**, 606 (1996)
D.N.C. Lin, S. Ida, ApJ **477**, 781 (1997)
J.J. Lissauer, D.C. Fabrycky, E.B. Ford et al., Nature **470**, 53 (2011a)
J.J. Lissauer, D. Ragozzine, D.C. Fabrycky et al., ApJS **197**, 8 (2011b)
J.J. Lissauer, D. Jontof-Hutter, J.F. Rowe et al., ApJ **770**, 131 (2013)
A. Loeb, B.S. Gaudi, ApJ **588**, L117 (2003)
E.D. Lopez, J.J. Fortney, ApJ **792**, 1 (2014)
S.H. Lubow, S. Ida, Exoplanets, ed. by S. Seager (University of Arizona Press, Tucson, AZ, 2011), pp. 347–371
K. Masuda, ApJ **783**, 53 (2014)
K. Masuda, Y. Suto, PASJ **68**, L5 (2016)
M. Mayor, D. Queloz, Nature **378**, 355 (1995)
J. Miralda-Escudé, ApJ **564**, 1019 (2002)
S.L. Morris, S.A. Naftilan, ApJ **419**, 344 (1993)

References

D. Naef, D.W. Latham, M. Mayor et al., A&A **375**, L27 (2001)
S.N. Raymond, C. Cossou, MNRAS **440**, L11 (2014)
L.A. Rogers, ApJ **801**, 41 (2015)
J. Sahlmann, P.F. Lazorenko, D. Ségransan et al., A&A **556**, A133 (2013)
J. Sahlmann, C. Lovis, D. Queloz, D. Ségransan, A&A **528**, L8 (2011)
R. Sanchis-Ojeda, D.C. Fabrycky, J.N. Winn et al., Nature **487**, 449 (2012)
H. Shibahashi, D.W. Kurtz, MNRAS **422**, 738 (2012)
A. Shporer, J.M., Jenkins, J.F., Rowe et al., AJ **142**, 195 (2011)
G. Takeda, F.A. Rasio, ApJ **627**, 1001 (2005)
S. Udry, M. Mayor, N.C. Santos, A&A **407**, 369 (2003)
J.E. Van Cleve, D.A. Caldwell, M.R. Haas, S.B. Howell (2016)
S.S. Vogt, R.P. Butler, E.J. Rivera et al., ApJ **787**, 97 (2014)
S. Weinberg, *Cosmology* (Oxford University Press, 2008)
L.M. Weiss, G.W. Marcy, ApJ **783**, L6 (2014)
L.M. Weiss, L.A. Rogers, H.T. Isaacson et al., ApJ **819**, 83 (2016)
A. Wolszczan, D.A. Frail, Nature **355**, 145 (1992)
J.T. Wright, G.W. Marcy, A.W. Howard et al., ApJ **753**, 160 (2012)
J.T. Wright, S. Upadhyay, G.W. Marcy et al., ApJ **693**, 1084 (2009)

Chapter 2
Measurements of Stellar Obliquities

Abstract As mentioned in Chap. 1, stars hosting hot Jupiters exhibit a wide range of obliquities. The knowledge comes from the spectroscopic technique, that is, modeling of the Rossiter-McLaughlin (RM) effect in the radial velocity (RV) times series. While the RM effect is difficult to observe for planets with smaller radii or on wider orbits than hot Jupiters, new methods based on the high-precision photometric data have recently been developed to provide complimentary information on those planets. In this chapter, we review the methods to measure stellar obliquities and summarize our current knowledge from observations. The implications for the formation scenario of hot Jupiters will be discussed in the next chapter.

Keywords Stellar obliquity · Spin–orbit angle · The Rossiter-Mclaughlin effect Photometric measurements of stellar obliquities

2.1 Definition and Terminology

The *stellar obliquity*, or the *spin–orbit angle*, is the angle between the stellar spin and planetary orbital axes, defined between 0 and π. Throughout this thesis, we use ψ to denote this angle. We call the orbits with $\psi < \pi/2$ *prograde*, and those with $\psi > \pi/2$ *retrograde*.

It is usually difficult to measure ψ for individual systems. Instead, it is easier to measure either one of the sky-plane or line-of-sight components of the true stellar obliquity ψ, which are illustrated in Fig. 2.1. The former angle, denoted by λ (Ohta et al. 2005), is called the *sky-projected* obliquity; it is the angle of the sky-projected orbital axis measured counter-clockwise from the sky-projected spin axis. The line-of-sight misalignment can be inferred from the *stellar inclination* $i_\star$, which is the direction of the stellar spin axis relative to our line of sight. For transiting exoplanets with their orbital inclination i_orb close to $\pi/2$, stellar inclination significantly different from $\pi/2$ immediately concludes the spin–orbit misalignment, while the opposite is not necessarily the case. Essentially, the two angles $i_\star$ and λ serve as the polar and azimuth angles to specify the direction of the stellar spin vector relative to the orbital one in three dimensions, with Z-axis being our line of sight. The true stellar

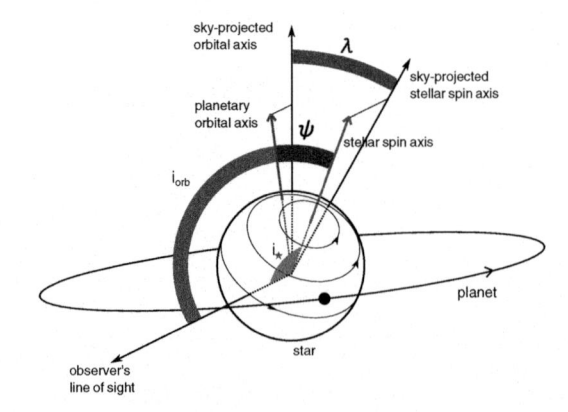

Fig. 2.1 Definitions of $i_{\rm orb}$, $i_\star$, λ, and ψ in this thesis. The orbital inclination, $i_{\rm orb}$, is the angle between the planetary orbital axis (blue arrow) and the observer's line of sight. In a transiting system, $i_{\rm orb}$ is usually very close to $\pi/2$ and hence the orbital axis almost coincides with its projection onto the plane of the sky. Inclination of the stellar spin axis, $i_\star$, is similarly defined as the angle between the stellar spin axis (red arrow) and the line of sight. The angle between the two axes (red and blue ones), ψ, is the stellar obliquity or the spin–orbit angle. Its sky projection, λ, denotes the angle between the sky projections of the same two axes

obliquity ψ is related to the sky-projected angle λ and the two inclinations $i_{\rm orb}$ and $i_\star$, via the law of cosines in spherical trigonometry:

$$\cos \psi = \cos i_{\rm orb} \cos i_\star + \sin i_{\rm orb} \sin i_\star \cos \lambda. \qquad (2.1)$$

Note that the measurements of obliquities discussed in this thesis are all for transiting systems. In fact, it is always advantageous to measure ψ in transiting systems, because $i_{\rm orb}$, one of the three angles required to specify ψ, is already fixed very precisely.

2.2 Obliquity from Spectroscopic Transit

Obliquities have traditionally been measured using the spectroscopic technique. The two techniques described here, both based on the same phenomenon, allow us to measure the sky-projected angle λ.

2.2.1 The Rossiter-McLaughlin Effect

Stellar rotation, which is usually faster than the planet-induced stellar motion by an order-of-magnitude, does not usually affect RVs. This is because the rotational velocity profile of a star is symmetric with respect to its sky-projected rotation axis;

2.2 Obliquity from Spectroscopic Transit

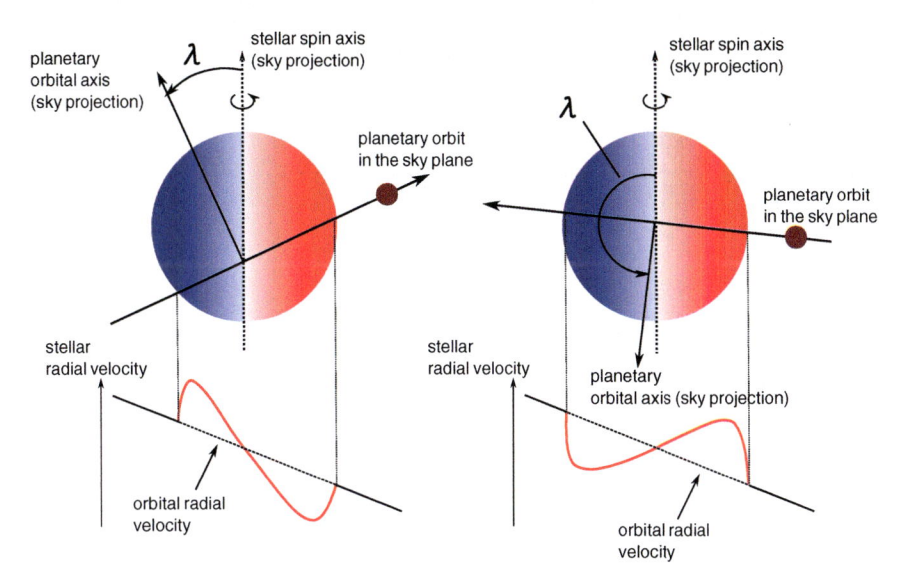

Fig. 2.2 Schematic illustration of the Rossiter-McLaughlin effect. The left panel illustrates a misaligned prograde orbit, while the right one shows a retrograde case

half of the surface is moving toward us, while the other away, and their net contribution is zero. In other words, stellar rotation only causes a symmetric broadening of its absorption lines, which does not shift the center of the lines.

A transiting planet breaks this symmetry and results in anomalous RV variations, known as the *Rossiter-McLaughlin effect* (Rossiter 1924; McLaughlin 1924). The pattern of the velocity anomaly depends on the relationship between the stellar rotation axis and orbit of the transiting planet, both projected onto the plane of the sky (see Fig. 2.2). If the planetary orbit is prograde, for example, the planet first blocks the approaching side of the star and then the receding side, and so the star apparently moves away, and then toward us (left panel). If the orbit is retrograde, on the other hand, the opposite pattern is observed (right panel).

As a first-order approximation, the RM effect can be simply described as a shift in the intensity-weighted centroid of the absorption lines in the velocity space (Ohta et al. 2005). Its amplitude $\Delta v_{\rm RM}$ is then given by

$$\Delta v_{\rm RM} = v \sin i_\star \left(\frac{R_{\rm p}}{R_\star}\right)^2 \sqrt{1-b^2} = 100\,{\rm m\,s^{-1}} \left(\frac{v \sin i_\star}{10\,{\rm km\,s^{-1}}}\right) \left(\frac{R_{\rm p}/R_{\rm Jup}}{R_\star/R_\odot}\right)^2 \sqrt{1-b^2}, \quad (2.2)$$

where $v \sin i_\star$ is the line-of-sight component of the stellar rotational velocity and b is the impact parameter of the transit normalized to $R_\star$ (cf. Appendix B). The formula gives a useful order-of-magnitude estimate for the expected anomaly, although we need to take into account other complicated effects for a more precise, quantitative analysis. Note that the value of $\Delta v_{\rm RM}$ is comparable to or even larger than that of the orbital RVs for a Jupiter-sized planet (cf. Eq. 1.4), while the detection is challenging

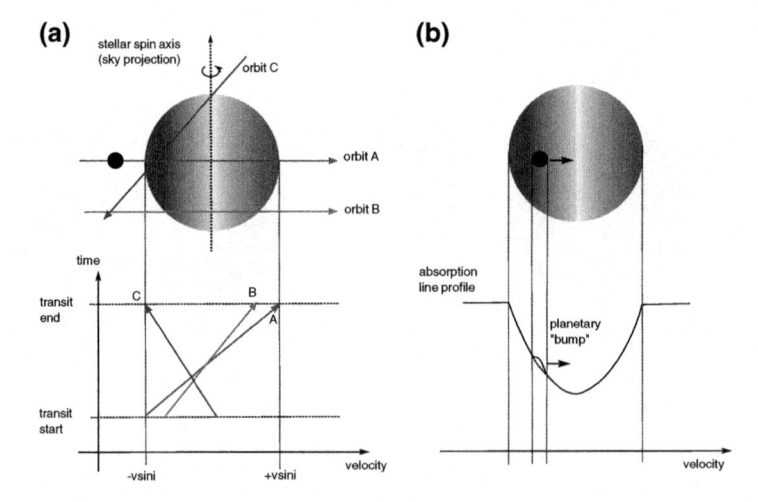

Fig. 2.3 Schematic illustration of the Doppler tomography method. **a** The bottom panel shows the motion of the planetary shadow (i.e., radial velocity corresponding to the central wavelength of the shadow) as a function of time, for the three different orbits illustrated in the top panel. **b** Illusrtation of the planetary shadow in the absorption line profile

for smaller planets and/or stars with a rotation velocity similar to the sun (about 2 km s^{-1}).

Precisely speaking, a transiting planet does not induce the *net* shift of the absorption lines, as is the case for the orbital motion. Rather, it *distorts* the line profile (cf. panel (b) of Fig. 2.3), and fitting such distorted lines with a symmetric template produces the anomalous velocity variations. The RM anomaly, therefore, depends on the specific manner to derive the velocity shift from given absorption lines and does not agree with the value computed as a centroid shift in general. Such deviations from the value computed with the centroid formula by Ohta et al. (2005) was first pointed out by Winn et al. (2005). The improved formulae taking into account specific procedures of the analysis, as well as other minor but significant effects to shape the absorption lines, have been developed by Hirano et al. (2010, 2011) for the iodine-cell technique, and by Boué et al. (2013) for the cross-correlation based method.

For a reliable measurement of the sky-projected obliquity λ with the RM effect, it is essential to determine the time when $\Delta v_{\rm RM}$ becomes zero relative to the central time of the transit. The two times are equal when $\lambda = 0°$, while in the case of the left panel in Fig. 2.2, the former time, when the planetary orbit crosses the sky-projected stellar rotation axis, is earlier than the central time of the transit, i.e., the midpoint of the transit chord. For this reason, a joint analysis with the photometric transit light curve (which determines the transit center), along with a reliable determination of the orbital radial velocity (which fixes the zero point of the RV anomaly), greatly improves the precision and accuracy of the measurement (Winn et al. 2005 and Chap. 4 of this thesis). It is also ideal that the transit impact parameter is not too close to zero, or the RM anomaly is always symmetric with respect to the transit center even for a spin–orbit misaligned case. The value of λ obtained in that case totally depends on the prior constraint on $v \sin i_\star$ and needs to be taken with care.

2.2.2 Doppler Tomography

As explained in the previous section, the essence of the RM effect is the distortion, rather than the shift, of the stellar absorption lines. The technique described here is to directly detect the distortion as a function of time. The shape of the line distortion is characterized as a "bump" in the absorption lines, whose position and width are determined by the line-of-sight rotation velocity distribution under the planetary disk, i.e., position and radius of the planetary disk (see panel (b) of Fig. 2.3). During a planetary transit, the central wavelength of the bump thus moves accordingly to the planetary motion. The range of wavelength/line-of-sight velocity over which the bump, or the planetary "shadow" moves, depends on the transit impact parameter and sky-projected obliquity λ, as illustrated in Fig. 2.3a. The method was first applied to a transiting planet by Collier Cameron et al. (2010a).

While it is more demanding to extract the subtle planetary shadow from the noisy spectra than to measure RVs, this technique, if applicable, allows for a more precise measurement of λ with fewer assumptions than the RM measurement, without the ambiguity of separating the orbital RVs and RM anomaly for instance (Albrecht et al. 2013). Moreover, it provides a unique possibility to constrain obliquities of fast-rotating planet-hosting stars, for which RVs (and hence the usual RM effect) cannot be measured very precisely due to the significant rotational broadening of the spectral lines (Collier Cameron et al. 2010b; Johnson et al. 2014; Bourrier et al. 2015). The same is also true for early-type stars that exhibit few absorption lines, for which precise RV velocimetry is impossible.

2.3 Obliquity from High-Precision Photometry

High-precision, continuous photometry as obtained by *Kepler* opened up new possibilities to gauge ψ. As will be described in detail below, they basically constrain $i_\star$ and are often applicable regardless of the planet properties. They are therefore complementary to the spectroscopic methods both in terms of applicable targets and derived information.

2.3.1 Asteroseismology

The long-term, uninterrupted, and extremely precise data of the stellar brightness provided by space-borne instruments, including *MOST* (Walker et al. 2003), *CoRoT* (Baglin et al. 2006a, b), and *Kepler* (Borucki et al. 2010), have made it possible to probe the internal structures of many stars through the detection of their oscillation modes with unprecedented precisions. The frequencies of oscillations, which are determined by the internal structure of the star (i.e., property of the cavity), provide precise knowledge about the stellar interior that is otherwise far out of reach. Such information from asteroseismology is valuable for precise and accurate characterization of explanetary systems as well (e.g., Carter et al. 2012), not to mention the stellar physics. More details on the recent development of asteroseismology may be found in recent conference proceedings (e.g., Shibahashi et al. 2012; Shibahashi and Lynas-Gray 2013; Guzik et al. 2014).

In addition to the fundamental stellar properties, asteroseismology also reveals the direction of the stellar rotation axis (i.e., stellar inclination $i_\star$ in Fig. 2.1) through the amplitudes (rather than frequencies) of the oscillation spectrum (Gizon and Solanki 2003). As is exactly the case for energy eigenstates of a quantum mechanical system in a spherically symmetric potential, each oscillation mode is labeled by three quantum numbers (n, l, m). While the $(2l+1)$-modes with the same (n, l) but different m have the same frequencies in the absence of stellar rotation, these degenerate modes can be separated in the power spectrum once the stellar rotation breaks the degeneracy, resulting in the typical splitting of $1/P_{\rm rot}$ with $P_{\rm rot}$ being the stellar rotation period. These split modes would usually have the same energy because we expect that the pressure-mode oscillation observed for Sun-like stars is excited by turbulence without any preferred direction. The disk-integrated strength[1] of each mode with different m, however, depends on the viewing angle; the modes with more angular nodes always visible from the observer tend to be weaker, because the oscillations on both sides of a node cancel each other. The relative heights of the modes with the same (n, l) but with different m thus yields $i_\star$ through the following simple formula:

$$\mathcal{E}(l, m, i_\star) = \frac{(l-|m|)!}{(l+|m|)!} \left[P_l^{|m|}(\cos i_\star) \right]^2, \quad (2.3)$$

where $P_l^{|m|}$ is the associated Legendre function, and $\mathcal{E}$ integrated over $0 < \cos i_\star < 1$ is normalized by $(2l+1)^{-1}$.

For transiting systems with orbital inclination close to $\pi/2$, stellar inclination gives a measure of the "line-of-sight" misalignment, which is complementary to the "sky-projected" misalignment constrained from the RM effect (see Fig. 2.1). In fact, if the stellar inclination $i_\star$ thus obtained is combined with the RM effect, true stellar obliquity ψ, which is usually hard to constrain, is obtained. In Chap. 4 we discuss the first attempt of such an application. For the planets not amenable to the RM measurement due to their small radii or long orbital periods, on the other hand, the constraint on ψ from asteroseismic $i_\star$ alone is usually not very strong for Sun-like stars. Rather it is better suited for statistical inference, given the advantage that the method can be applied regardless of the property of the planets (Campante et al. 2016). Meanwhile, $i_\star$ can be constrained more precisely for evolved stars, for which asteroseismology led to an important discovery (Huber et al. 2013, see Sect. 2.4.3).

---

[1]Remember that the surface is not resolved for stars, unlike the case of asteroseismology for the sun (i.e., helioseismology).

2.3.2 Gravity Darkening

Centrifugal force due to rapid stellar rotation reduces the surface gravity around the stellar equator and elongates it relative to the pole. The elongation expands the intervals of the equipotential surfaces, and hence reduces the temperature gradient. As a result, stellar flux becomes smaller around the equator than the pole. This is the phenomenon known as *gravity darkening*, and the flux dependence on the surface gravity g is given by

$$T_{\text{eff}} \propto g^{\beta}, \quad \beta = 0.25 \tag{2.4}$$

for a star with a radiative envelope (von Zeipel's law; von Zeipel 1924).

If a planet transits such a "gravity-darkened" star, the above equator-to-pole brightness contrast deforms the transit light curve (Fig. 2.4). Since the shape of this anomaly depends on the position of the bright stellar pole relative to the planetary orbit, stellar obliquity can be inferred from the gravity-darkened transit light curve (Barnes 2009). The method has been applied to several transiting planets (Barnes et al. 2011, 2015; Ahlers et al. 2015) and eclipsing binaries around fast-rotating stars (Philippov and Rafikov 2013; Zhou and Huang 2013; Ahlers et al. 2014), for which obliquity measurements with other techniques are challenging. In the only case where the Doppler tomography was also applicable (Johnson et al. 2014), however, the result from the gravity-darkening method was shown to disagree with the latter measurement. Moreover, gravity-darkening measurements performed by different authors sometimes report inconsistent results (Zhou and Huang 2013; Ahlers et al. 2014) for some unknown reason. These issues will be further discussed in Chap. 5.

The theory of gravity darkening has been directly tested by imaging the surface brightness distribution of nearby rapid rotators with interferometry (e.g. Monnier et al. 2007), and more indirectly with the modeling of ellipsoidal variations of close

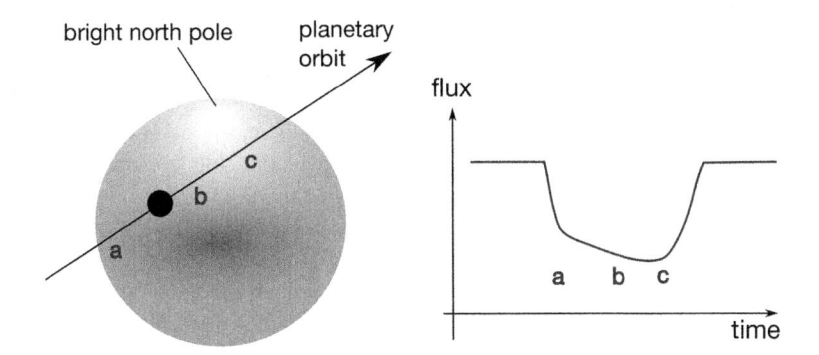

Fig. 2.4 Schematic illustration of the gravity-darkened transit. The transit is deepest when the planet is closest to the bright pole of the star (white region). Note that here we only show the brightness distribution due to the gravity darkening, while the actual brightness profile is dominated by the limb darkening

binaries (e.g. Djurašević et al. 2006). While the observed profile largely agrees with the theoretical predictions, some observations report possible deviation from the classical von Zeipel law. This may be due to the very rapid rotation close to the break up (Espinosa Lara and Rieutord 2011) or may be due to the poorly-understood processes including the convection or magnetic field (Rieutord 2015). Indeed, gravity darkening of lower-mass stars with convective envelope seems far from being understood.

2.3.3 Spectroscopic $v \sin i_\star$ and Stellar Rotation Period

The width of absorption lines yield $v \sin i_\star$, the projected rotational velocity of the star. If we have the independent knowledge on $v = 2\pi R_\star / P_{\rm rot}$, the stellar equatorial rotation velocity, we can constrain $i_\star$ and compare it to $i_{\rm orb} \simeq \pi/2$ for transiting systems. The stellar radius can usually be estimated by stellar modeling based on the stellar atmospheric parameters obtained from spectroscopy, or by asteroseismology if applicable. The problem is how to estimate $P_{\rm rot}$, for which two methods have been proposed.

The first is to rely on the empirical relation between the stellar age and the rotation period, which forms the basis of gyrochronology. In general, older stars tend to rotate more slowly presumably due to the magnetic braking, which produces a good correlation with the age and rotation period. While gyrochronology estimates the stellar age from the rotation period, we could use the relation in an opposite way to obtain the latter from the former, derived from spectroscopy or asteroseismology. Schlaufman (2010) applied this method to 75 transiting planets and identified 10 systems exhibiting possible spin–orbit misalignments.

The second is to use photometric modulation of the star due to the star spots on the stellar surface. The *Kepler* data made it possible to infer the rotation periods of tens of thousands of stars in this way (McQuillan et al. 2013; Walkowicz and Basri 2013; McQuillan et al. 2014). Hirano et al. (2012a, 2014) applied the method to ∼100 transiting systems for which spectroscopic $v \sin i_\star$ and $R_\star$ are obtained. While they did not find any significant difference for single- and multi-transiting samples, Morton and Winn (2014), who adopted a more sophisticated statistical approach, presented a piece of evidence that planets in single-transiting systems may have higher obliquities than those in multi-transiting systems.

Both methods do not give very strong constraints on individual systems due to relatively large uncertainties in the spectroscopic $v \sin i_\star$ and $R_\star$. In addition, in the first method the rotation period estimated from the gyrochronological relation is very uncertain due to the inherent scatter in the empirical relation as well as the difficulty in precisely estimating the stellar age. Nevertheless, this method can be applied to a large number of systems and thus may benefit the statistical inference with a large sample.

2.3.4 Spot Anomaly

If a transiting planet crosses over a star spot, we observe an instantaneous increase in the relative flux because of the smaller intensity within the spot. Such anomalies, if observed multiple times via continuous photometric monitoring, also allow for constraining stellar obliquities (Sanchis-Ojeda et al. 2011). While it is a priori likely to observe such recursive spot crossings for lower-obliquity systems, the method is also applicable, at least in principle, to the misaligned case as demonstrated by Sanchis-Ojeda and Winn (2011).

In addition to the in-transit anomaly, star spots also induce out-of-transit flux modulation over the timescale of stellar rotation period. The modulation, if continuously monitored as well, greatly helps the above decoding, because the global modeling of such out-of-flux modulations reveals the stellar rotation period and even the longitudinal phase of the spot at a given epoch (Nutzman et al. 2011; Sanchis-Ojeda et al. 2012). The essentially same effect also manifests as the correlation between the local derivative of the out-of-transit flux and shifts in the transit times induced by the spot anomaly; the correlation can be used to distinguish the prograde and retrograde motions (Mazeh et al. 2015a).

2.3.5 Spot-Modulation Amplitude

The methods discussed so far are, at least in principle, applicable to individual systems, while this method is statistical in nature.

Brightness modulation due to star spots tend to be weaker when the stars are seen from the pole. For such a configuration, star spots around the stellar equator, as observed for the sun, are located close to the limb of the stellar disk, and so they produce only minor modulations due to geometric foreshortening and limb darkening. Thus, if we compare the spot-modulation amplitudes of stars that host transiting planets (where planetary orbits are close to edge on) with that do not, we can evaluate correlation between the stellar inclination and planetary orbital inclination: the stars hosting transiting planets should show the modulations of larger amplitudes than those without transiting planets, if the stellar equatorial plane tends to be aligned with the planetary orbital plane.

Mazeh et al. (2015b) applied this analysis to 993 KOIs (i.e., stars hosting candidate transiting planets) and 33614 *Kepler* stars with no known transiting planets. They found that cool planet-hosting stars with $T_{\rm eff} \lesssim 6000$ K exhibit a clear signature of the spin–orbit correlation (i.e., alignment), while their hotter counterparts show a weaker correlation and hence likely have higher stellar obliquities, assuming that their spot distribution is similar to that of the cooler stars. Since the majority of these KOIs are relatively small planets on wider orbits than hot Jupiters, the result demonstrates that the high obliquity of hot stars identified in the RM sample is not specific to hot Jupiters, as will be discussed in Sect. 2.4.1.

More importantly, they found no sharp obliquity dependence on the orbital period for the cooler KOI sample, as later confirmed by Li & Winn (2016) with a more elaborated analysis.[2] The fact may argue against the scenarios that the current obliquity distribution has been sculpted by the tidal star–planet interaction, as will be discussed in the next chapter.

2.4 Correlations with the System Properties

As of April 2016, stellar obliquities have been measured for about 80 individual systems.[3] Since most of these constraints are from the spectroscopic transit observations, they are mainly Jupiter-sized planets and only the sky-projected obliquities λ are constrained. As we have seen in Fig. 1.5, λ of exoplanetary systems distribute broadly; about one third of the sample exhibit spin–orbit misalignments in terms of λ at the three-sigma level.

In this section, we summarize our current knowledge on the obliquity, both on λ from the RM measurements and statistical results from various photometric techniques described above. We especially focus on the correlation with other properties of the system, which will be an important clue to understand the origin of the spin–orbit misalignment. The theoretical interpretation will be separately discussed in Chap. 3.

2.4.1 Hot Stars (with Hot Jupiters) Have High Obliquities

The most significant trend in the stellar obliquity known to date is the correlation between the misalignment and the effective temperature of the host star (or whatever else correlated to the latter). Winn et al. (2010) first pointed out that large spin–orbit misalignments are preferentially found around hot stars with $T_{\text{eff}} \gtrsim 6250\,\text{K}$, which was confirmed by Albrecht et al. (2012) with a larger sample. Similar trends can be seen in terms of the stellar mass (Schlaufman 2010), stellar age (Triaud 2011), and stellar rotation period (Dawson 2014), which are all well correlated with the stellar effective temperature for main-sequence stars.

Figure 2.5 shows the updated compilation of this λ–T_{eff} relation for the same sample as in Fig. 1.5. Here we choose $T_{\text{eff}} = 6100\,\text{K}$ as a dividing line following Winn and Fabrycky (2015), who also discussed the latest statistics. The trend is still clear except for the four systems in the upper left part labeled with the planet names. It is worth noting that they all have relatively large $a/R_\star > 10$, and often have smaller masses (Fig. 2.6), i.e., they are not hot Jupiters. These features support the tidal origin of the trend, as will be discussed in the next chapter.

---

[2]Note that they did find a weak period dependence; see Sect. 3.2.1 for its implication.

[3]See, e.g., Holt-Rossiter-McLaughlin Encyclopaedia at http://www2.mps.mpg.de/homes/heller/.

2.4 Correlations with the System Properties

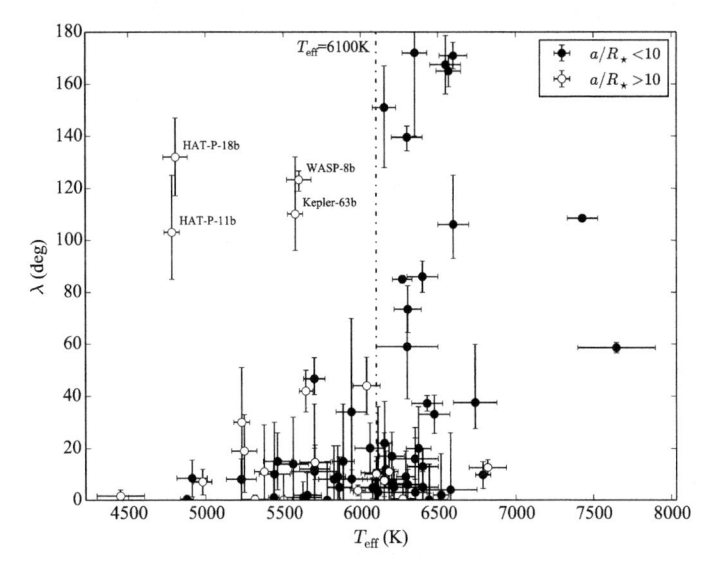

Fig. 2.5 The values of λ as a function of effective temperatures of the host stars. The vertical dashed line corresponds to $T_{\text{eff}} = 6100$ K. Filled circles denote close-in planets with $a/R_\star < 10$, while planets shown by open ones orbit farther away from the star ($a/R_\star > 10$)

Until recently, the trend has been discussed mainly in the context of hot Jupiter formation, as a natural consequence that obliquities have been measured only for hot Jupiters. The statistical inference using the spot-modulation amplitude (Sect. 2.3.5), however, recently showed that planets other than hot Jupiters are also likely to have higher obliquities around hotter stars (Mazeh et al. 2015b). While the result may indicate that the large spin–orbit misalignment is not specific to hot Jupiters and their formation process, the interpretation is quite uncertain at this point. The statistical nature of the spot-amplitude analysis only allows for the *relative* comparison between hot and cool stars, and so it is difficult to quantitatively assess how the "higher" obliquity found for planets around hotter stars compare to the high obliquity observed for hot Jupiters. It may also be possible that the surface distribution of spots on hotter stars is different from that on cooler stars, in which case the relationship between the spot-modulation amplitude and stellar obliquity would not be straightforward.

2.4.2 Planetary Mass Cut Off for Retrograde Planets

Discovery of planets on retrograde ($\psi > \pi/2$) orbits (e.g. Winn et al. 2009) was one of the most surprising outcomes of the RM measurements. Hébrard et al. (2011) pointed out that such retrograde orbits are only found for hot Jupiters less massive than $\sim 3 M_{\text{Jup}}$, as shown in Fig. 2.6. It may also be worth noting that the massive hot Jupiters on prograde orbits are mostly found around hot stars, around which

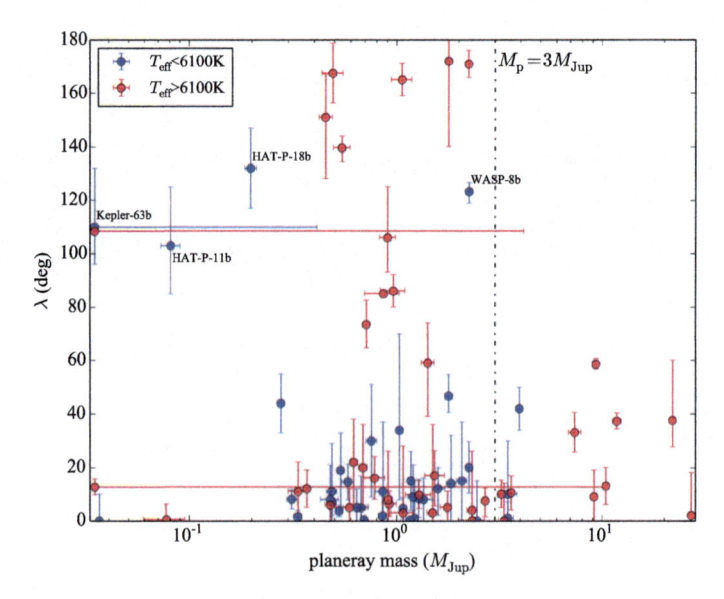

Fig. 2.6 The values of λ as a function of the planetary mass. The red and blue circles correspond to planets around hot ($T_{\text{eff}} > 6100$ K) and cool ($T_{\text{eff}} < 6100$ K) stars, respectively, and the vertical dashed line corresponds to $M_p = 3\,M_{\text{Jup}}$. Note that masses of the planets in this plot are not the minimum mass, since they are all transiting (or the RM effect cannot be measured). For clarity, planetary masses are set to the minimum value of the horizontal axis when only the upper limit is obtained

spin–orbit misalignments are more frequent. The fact may point to the effect of tidal star–planet interaction, whose strength increases proportionally to the planet-to-star mass ratio. Alternatively, it may simply suggest that these "super-Jupiter" mass objects were formed in a different manner from hot Jupiters with $M_p \lesssim 3\,M_{\text{Jup}}$.

2.4.3 Single- Versus Multi-transiting Systems

So far the spin–orbit misalignment is "rare" among multi-transiting systems. This supports the idea that the initial star–disk alignment as expected for the solar system is common, because the planets in multi-transiting systems presumably have well-aligned orbits and thus trace the plane of their natal protoplanetary disk.

Either the sky-projected obliquity or stellar inclination has been constrained for seven multi-transiting systems. Among these, only one system, Kepler-56, exhibits a clear spin–orbit misalignment ($i_\star = 47° \pm 6°$ from asteroseismology by Huber et al. 2013), while the other six are consistent with the alignment at least in terms of the sky-projected or line-of-sight component. The spectroscopic transits (Sect. 2.2) have been observed for Kepler-89d ($\lambda = -6°^{+13°}_{-11°}$ and $-11° \pm 11°$ by Hirano et al. 2012b, Albrecht et al. 2013 respectively), Kepler-25c ($\lambda = 7° \pm 8°$ by Albrecht et al. 2013), and for WASP-47b ($\lambda = 0° \pm 24°$ by Sanchis-Ojeda et al. 2015); asteroseismolgy (Chaplin et al. 2013, see also Sect. 2.3.1) points to alignments for Kepler-50 ($i_\star = 82°^{+8°}_{-7°}$) and Kepler-65 ($i_\star = 81°^{+9°}_{-16°}$); and $\lambda \lesssim 10°$ is obtained for Kepler-30 (Sanchis-Ojeda et al. 2012) from the spot anomaly (Sect. 2.3.4). In Chap. 4, we will report on a new measurement of true obliquity ψ, rather than λ, for Kepler-25c.

Statistical inferences for the line-of-sight misalignments (i.e., difference between $i_\star$ and i_{orb}; see Sects. 2.3.3 and 2.3.1) also support the alignment in multi-transiting systems, though only in a relative sense. Morton and Winn (2014) analyzed the sample of $v \sin i_\star$, $R_\star$, and P_{rot} for 70 KOIs using a hierarchical Bayesian technique and found marginal evidence that stars hosting single-transiting planetary systems have systematically higher obliquities than those hosting multi-transiting systems. The conclusion was further strengthened by adding asteroseismic samples (Campante et al. 2016). The possible difference between single- and multi-transiting systems, if real, does not only support the initial star–disk alignment, but also suggests that (a part of) excess single-transiting systems as implied by the multiplicity statistics (known as the *Kepler* dichotomy, e.g., Lissauer et al. 2011; Ballard and Johnson 2016) might actually represent the dynamically "hotter" (i.e., mutually more inclined) multi-planetary systems.

References

J.P. Ahlers, J.W. Barnes, R. Barnes, ApJ **814**, 67 (2015)
J.P. Ahlers, S.A. Seubert, J.W. Barnes, ApJ **786**, 131 (2014)
S. Albrecht, J.N. Winn, G.W. Marcy, et al., ApJ **771**, 11(A13) (2013)
S. Albrecht, J.N. Winn, J.A. Johnson, et al., ApJ **757**, 18(A12) (2012)
A. Baglin, M. Auvergne, P. Barge, et al., in *The Corot Mission Pre-launch Status-Stellar Seismology And Planet Finding*, ed. by M. Fridlund, A. Baglin, J. Lochard, L. Conroy, vol. 1306 (ESA Special Publication, 2006a), p. 33
A. Baglin, M. Auvergne, L. Boisnard, et al., in *COSPAR Meeting, 36th COSPAR Scientific Assembly*, vol. 36 (2006b)
S. Ballard, J.A. Johnson, ApJ **816**, 66 (2016)
J.W. Barnes, ApJ **705**, 683 (2009)
J.W. Barnes, J.P. Ahlers, S.A. Seubert, H.M. Relles, ApJ **808**, L38 (2015)
J.W. Barnes, E. Linscott, A. Shporer, ApJS **197**, 10(B11) (2011)
W.J. Borucki, D. Koch, G. Basri et al., Science **327**, 977 (2010)
G. Boué, M. Montalto, I. Boisse, M. Oshagh, N.C. Santos, A&A **550**, A53 (2013)
V. Bourrier, A. Lecavelier des Etangs, G. Hébrard et al., A&A **579**, A55 (2015)
T.L. Campante, M.N. Lund, J.S. Kuszlewicz et al., ApJ **819**, 85 (2016)
J.A. Carter, E. Agol, W.J. Chaplin et al., Science **337**, 556 (2012)
W.J. Chaplin, R. Sanchis-Ojeda, T.L. Campante et al., ApJ **766**, 101 (2013)
A. Collier Cameron, V.A. Bruce, G.R.M. Miller, A.H.M.J. Triaud, D. Queloz, MNRAS **403**, 151 (2010a)
A. Collier Cameron, E. Guenther, B. Smalley et al., MNRAS **407**, 507 (2010b)
R.I. Dawson, ApJ **790**, L31 (2014)
G. Djurašević, H. Rovithis-Livaniou, P. Rovithis et al., A&A **445**, 291 (2006)
F. Espinosa Lara, M. Rieutord, A&A **533**, A43 (2011)

L. Gizon, S.K. Solanki, ApJ **589**, 1009 (2003)
J.A. Guzik, W.J. Chaplin, G. Handler, A. Pigulski, (eds.), in *IAU Symposium, Precision Asteroseismology*, vol. 301 (2014)
G. Hébrard, D. Ehrenreich, F. Bouchy et al., A&A **527**, L11 (2011)
T. Hirano, R. Sanchis-Ojeda, Y. Takeda et al., ApJ **756**, 66 (2012a)
T. Hirano, R. Sanchis-Ojeda, Y. Takeda et al., ApJ **783**, 9 (2014)
T. Hirano, Y. Suto, A. Taruya et al., ApJ **709**, 458 (2010)
T. Hirano, Y. Suto, J.N. Winn et al., ApJ **742**, 69 (2011)
T. Hirano, N. Narita, B. Sato et al., ApJ **759**, L36 (2012b)
D. Huber, J.A. Carter, M. Barbieri et al., Science **342**, 331 (2013)
M.C. Johnson, W.D. Cochran, S. Albrecht et al., ApJ **790**, 30 (2014)
G. Li, J.N. Winn, ApJ **818**, 5 (2016)
J.J. Lissauer, D. Ragozzine, D.C. Fabrycky et al., ApJS **197**, 8 (2011)
T. Mazeh, T. Holczer, A. Shporer, ApJ **800**, 142 (2015a)
T. Mazeh, H.B. Perets, A. McQuillan, E.S. Goldstein, ApJ **801**, 3 (2015b)
D.B. McLaughlin, ApJ **60**, 22 (1924)
A. McQuillan, T. Mazeh, S. Aigrain, ApJ **775**, L11 (2013)
A. McQuillan, T. Mazeh, S. Aigrain, ApJS **211**, 24 (2014)
J.D. Monnier, M. Zhao, E. Pedretti et al., Science **317**, 342 (2007)
T.D. Morton, J.N. Winn, ApJ **796**, 47 (2014)
P.A. Nutzman, D.C. Fabrycky, J.J. Fortney, ApJ **740**, L10 (2011)
Y. Ohta, A. Taruya, Y. Suto, ApJ **622**, 1118 (2005)
A.A. Philippov, R.R. Rafikov, ApJ **768**, 112 (2013)
M. Rieutord (205), arXiv:1505.03997
R.A. Rossiter, ApJ **60**, 15 (1924)
R. Sanchis-Ojeda, J.N. Winn, ApJ **743**, 61 (2011)
R. Sanchis-Ojeda, J.N. Winn, M.J. Holman et al., ApJ **733**, 127 (2011)
R. Sanchis-Ojeda, D.C. Fabrycky, J.N. Winn et al., Nature **487**, 449 (2012)
R. Sanchis-Ojeda, J.N. Winn, F. Dai et al., ApJ **812**, L11 (2015)
K.C. Schlaufman, ApJ **719**, 602 (2010)
H. Shibahashi, A.E. Lynas-Gray (eds.), *Progress in Physics of the Sun and Stars*, Astronomical Society of the Pacific Conference Series (2013)
H. Shibahashi, M. Takata, A.E. Lynas-Gray (eds.), *Progress in Solar/Stellar Physics with Helio- and Asteroseismology*, Astronomical Society of the Pacific Conference Series, vol. 462 (2012)
A.H.M.J. Triaud, A&A **534**, L6 (2011)
H. von Zeipel, MNRAS **84**, 665 (1924)
G. Walker, J. Matthews, R. Kuschnig et al., PASP **115**, 1023 (2003)
L.M. Walkowicz, G.S. Basri, MNRAS **436**, 1883 (2013)
J.N. Winn, D. Fabrycky, S. Albrecht, J.A. Johnson, ApJ **718**, L145 (2010)
J.N. Winn, D.C. Fabrycky, ARA&A **53**, 409 (2015)
J.N. Winn, J.A. Johnson, S. Albrecht, et al., ApJ **703**, L99(W09) (2009)
J.N. Winn, R.W. Noyes, M.J. Holman et al., ApJ **631**, 1215 (2005)
G. Zhou, C.X. Huang, ApJ **776**, L35 (2013)

Chapter 3
Origin of the Misaligned Hot Jupiters: Nature or Nurture?

Abstract What produces the spin–orbit misalignment of hot Jupiters? One natural speculation would be that it is related to the specific formation channel of hot Jupiters, i.e., their orbital migration. The "high-eccentricity migration" scenario, tidal migration following the eccentricity excitation through few-body dynamical processes, can naturally produce the spin–orbit misalignment along with the highly eccentric orbits as mentioned in Sect. 1.1.4. In Sect. 3.1, we describe this scenario in detail. The challenge to this (and actually to any other) scenario is the correlation between the obliquity and effective temperature of hot-Jupiter hosts discussed in Sect. 2.4.1. A possible explanation is that the trend is attributed to the different timescales for obliquity damping in cool and hot stars (Sect. 3.2). The subsequent studies, however, show that it is difficult to reproduce the trend at least with the current theory of tides. In addition, evidence against this "tidal realignment" scenario has recently been presented by new measurements of obliquities with the *Kepler* photometry. It has also been pointed out that it may be difficult to produce the most misaligned hot Jupiters within this framework of migration. Given the situation, another class of scenarios without resorting to the violent dynamical events has also been proposed; this is the topic of Sect. 3.3. These scenarios consider the misalignment to be of "primordial" origin, that is, the misalignment between the stellar spin and protoplanetary disk. They might consistently explain the obliquity dependence on the host star, as well as the presence of counter-orbiting hot Jupiters. In this chapter, we review both of these "nature" and "nurture" scenarios along with their strengths and weaknesses, and discuss how the trend could be explained in each class of scenarios. We also propose several important questions that need to be addressed to distinguish the two scenarios.

Keywords High-eccentricity migration · Hot Jupiter · Star–disk misalignment
Tidal star–planet interactions

3.1 High-Eccentricity Migration

This class of migration scenarios was proposed right after the first discovery of a hot Jupiter and the proposal of disk migration scenario to explain its close-in orbit (Lin et al. 1996). While this alternative was originally motivated by large eccentricities observed for relatively short-period Jupiters discovered in the early days (cf. Figs. 1.3 and 1.4), the frequent spin–orbit misalignments observed for transiting hot Jupiters later provided further support for this type of scenario.

3.1.1 The Scenario

While a variety of scenarios has been proposed for the high-eccentricity migration, all of them consist of the common two processes described below.

First, the planetary orbit (usually assumed to be beyond ~AU initially) acquires a large eccentricity close to unity via some dynamical process. This causes the planet to have a very small pericenter distance $a(1-e)$, which will eventually be comparable to the final semi-major axis of the resulting hot Jupiter.[1] For example, the eccentricity as large as ~0.99 is required for a Jupiter at $a = 5$ AU to migrate to $a = 0.05$ AU via this process. It is usually during this process that the spin–orbit misalignment is produced, because the process that excites eccentricity often involves the excitation of orbital inclination that drives the orbit out of the original disk plane. The timescale and condition for the eccentricity/inclination excitation significantly depends on the specific scenario.

Next, tidal interaction enhanced around the close pericenter shrinks and circularizes the orbit. During a close encounter around the pericenter, tidal force from the central star distorts the planet and excites its oscillation, whose energy is eventually dissipated inside the planet. While the energy dissipation reduces the orbital semi-major axis of the planet, the conservation of orbital angular momentum,[2] which is proportional to $\sqrt{a(1-e^2)}$ (cf. Appendix A), requires that e is also reduced, i.e., the orbit is circularized as well.

The final semi-major axis of the circularized hot Jupiter, a_f, is thus simply related to the pericenter distance at the onset of circularization, q_c, via

$$a_f = a(1-e^2) = q_c(1+e_c) \simeq 2q_c, \qquad (3.1)$$

where $e_c \sim 1$ at the beginning of circularization. This implies that the semi-major axis distribution of hot Jupiters formed via high-eccentricity migration should have

---

[1] As we will see below, the final semi-major axis is actually *twice* the pericenter distance.

[2] Assuming that the planetary spin is already synchronized with the orbit; this usually occurs on a much shorter timescale than the orbit circularization by the ratio of the planetary moment of inertia to $M_p a^2$ (Correia 2009).

an inner edge twice the Roche limit a_Roche of the central star. Rasio and Ford (1996) actually showed that the observational data favor $2a_\text{Roche}$ rather than a_Roche as the inner edge.

Below we describe several possible mechanisms for the eccentricity excitation in the first step. Note that these mechanisms are not necessarily mutually exclusive.

3.1.1.1 Secular Interaction with a Stellar and Planetary Companion: Kozai–Lidov Cycles with Tidal Friction

Let us consider a hierarchical three-body astrophysical system, where the inner semi-major axis, a_in, is much smaller than that of the outer orbit, a_out.[3] Even the weak gravitational perturbation from such a distant outer object can gradually accumulate to affect the long-term behavior of the inner orbit. Kozai (1962) found that, if the inner orbit is inclined with respect to the outer one by more than $i_\text{crit} \sim 40°$, the inner orbit, even if initially circular, experiences the periodic excursion of orbital eccentricity to a large value coupled with the oscillation of orbital inclination. This is known as the Kozai cycle.

Because the timescale of such an evolution is much longer than the orbital one, the long-term behavior of the system can be tracked by considering the potential averaged over the inner and outer orbits, i.e., interaction between the two rigid "rings" (Gauss's method; Murray and Dermott 1999). In this "secular" approximation, the orbital semi-major axis (i.e., orbital energy) is conserved because the potential is time-independent. In addition, to the lowest order in a_in/a_out (quadrupole approximation), the Kozai integral

$$H = \sqrt{1 - e^2} \cos i \tag{3.2}$$

is conserved during the eccentricity/inclination oscillation. This relation represents the conservation of the semi-major axes and the angular momentum normal to the outer orbit, the latter of which follows from the axisymmetry of the potential due to the outer companion. Thus, if the value of H is sufficiently small (i.e., $\cos i$ was small when the orbit was initially circular), the inner orbit can acquire a large eccentricity (or a small $1 - e^2$) when i becomes small in the cycle. The timescale for the oscillation is given by

$$P_\text{Kozai} = \frac{2}{3\pi} \frac{m_\text{tot}}{m_\text{out}} \frac{P_\text{out}^2}{P_\text{in}} (1 - e_\text{out}^2)^{3/2}, \tag{3.3}$$

where m_tot and m_out are the masses of the whole system and outer object, respectively (Kiseleva et al. 1998), and $P_\text{in/out}$ are the orbital periods with their subscripts denoting

---

[3] For multi-stellar systems, such a hierarchy is a natural consequence of the requirement of dynamical stability, while it is not necessarily the case for two-planet systems.

the inner and outer orbits. The timescale is about 10^7 yr for a stellar perturber at $a_\text{out} \sim 1000$ AU, and depends a lot on the system parameters; note the strong a_out^3 dependence.

When combined with the tidal dissipation, this provides a natural mechanism for producing the short-period binaries (Mazeh and Shaham 1979), which is the scenario known as "Kozai cycles with tidal friction" (KCTF, Kiseleva et al. 1998; Eggleton and Kiseleva-Eggleton 2001). Also for planetary systems, there are at least two examples known for which the Kozai cycle is likely responsible for the observed high orbital eccentricity (Holman et al. 1997; Wu and Murray 2003), although the whole eccentricity distribution cannot be explained by the Kozai mechanism alone (Takedo and Rasio 2005). Comprehensive studies of hot Jupiter formation via this mechanism have been performed by Fabrycky and Tremaine (2007) and Wu et al. (2007), who made predictions for the obliquity distributions of hot Jupiters produced in this mechanism.

While the above studies consider the stellar object as an outer perturber, a similar migration can also occur in a hierarchical two-planet system, where the outer perturber is a planet (Naoz et al. 2011). In such a case, the Kozai integral (i.e., the angular momentum normal to the invariant plane) is not necessarily constant due to the higher-order terms of secular perturbation (e.g., Ford et al. 2000), and so the inner orbit can even flip its direction. Such a higher-order term (mainly octupole) can also play a significant role when the outer orbit is eccentric, in which case the large mutual inclination is not necessarily required to produce a large inner eccentricity (Li et al. 2014). That is, high-eccentricity migration is also possible for a coplanar system (Petrovich 2015), where the resulting close-in planet can be spin–orbit aligned. Although this mechanism has been proposed as a viable path to produce "counter-orbiting" planets with $\psi \approx 180°$, (e.g., Naoz et al. 2011; Li et al. 2014), the most common outcome seems to be the tidal disruption (Xue and Suto 2016). In addition to those higher-order term effects, it has also been pointed out that the torque due to the rotationally-deformed quadrupole moment of the host star may add a further complexity in the obliquity evolution (Storch et al. 2014).

3.1.1.2 Planet–Planet Scattering

When more than one giant planets are formed on sufficiently close orbits, the long-term gravitational interaction can lead to the dynamical instability, which results in the close encounter and scattering between the planets. Some of the scattered planets acquire sufficiently large eccentricities for their pericenter distances to be close enough to tidally migrate. The process often involves an ejection or excitation of the outer planet's eccentricity, as well as excitation of the orbital inclination (i.e., spin–orbit misalignment).

3.1 High-Eccentricity Migration

While the semi-major axis after the scattering can only be about half of the original value for the two-planet case (Rasio and Ford 1996),[4] the outcome can be even more diverse if three or more giant planets are involved (Weidenschilling and Marzari 1996). The typical outcome in the three-giant case is the ejection of one planet, while the two survivors are left in well separated orbits one closer to the star and the other farther away, often with significant eccentricities and a large mutual inclination (Marzari and Weidenschilling 2002).

Even though the eccentricity of the inner planet does not reach a sufficiently large value after one scattering, the secular interaction (i.e., the Kozai effect) due to the scattered outer planet can further excite the innermost planet's eccentricity to enhance the chance of hot Jupiter formation. Especially, Nagasawa et al. (2008) pointed out that the repeated Kozai cycles during the three-planet orbit crossing, rather than the typical case of the two-survived planets, significantly contribute to the eccentricity excitation and increase the formation probability of close-in orbits by a factor of a few compared to the previous estimates (Marzari and Weidenschilling 2002; Chatterjee et al. 2008).

3.1.1.3 Secular Chaos

Wu and Lithwick (2011) showed that secular interaction between two or more well-spaced planets can lead to the chaotic diffusion of the eccentricity and inclination of the innermost planet. While the process requires eccentricities/inclinations of $O(1)$ for a system with two planets, the threshold is much reduced for a system with three or more planets, especially if the inner planet is the least massive. The scenario could in principle produce retrograde hot Jupiters depending on the (largely unknown) initial orbits of the system. It also predicts the presence of companions outside a few AU of hot Jupiters, as well as the rise in their frequency with increasing stellar age.

3.1.2 Relevant Observational Issues

How does the high-eccentricity migration scenario compare to observations? So far, observations seem to present both positive and negative results, as summarized below.

3.1.2.1 Existence of Highly Eccentric Planets

As mentioned in Sect. 1.1.4, the shear existence of highly eccentric planets seem to argue for the past dynamical events at least in some systems. Indeed, the observed

---

[4] When all the planets have equal masses, energy conservation requires that the post-scattering semi-major axis of the innermost planet, $a_{\text{final,in}}$, is bounded by $1/a_{\text{final,in}} = \sum_{j=1}^{N} 1/a_{\text{initial,j}} < N/\min(a_{\text{initial,j}})$ or $a_{\text{final,in}} > \min(a_{\text{initial,j}})/N$ (Nagasawa et al. 2008).

eccentricity distribution is well explained by the planet–planet scattering for the sample with $e \gtrsim 0.2$, largely regardless of the system configuration prior to the scattering phase (Chatterjee et al. 2008; Jurić and Tremaine 2008). The fact implies that the dynamical instability has played a universal role in sculpting the observed architecture of exoplanetary systems.

3.1.2.2 Three-Day Pile up of Hot Jupiters

Radial velocity surveys reported a "pile-up" of Jupiter-sized planets around $P = 3$ days in their log-period distribution (Cumming et al. 1999; Udry et al. 2003). While the presence of this pile-up was called into question by the following studies of the *Kepler* data (e.g., Howard et al. 2012), Dawson and Murray-Clay (2013) found that the peak is recovered even in the *Kepler* sample if only the samples with super-solar metallicities are considered. In addition, comprehensive RV observations of the *Kepler* giant planets recently reported by Santerne et al. (2016) also confirmed the three-day pile-up. Indeed, this pile-up is the very feature expected from the high-eccentricity migration scenario (Wu et al. 2007; Fabrycky and Tremaine 2007; Wu and Lithwick 2011), except for the case where a high eccentricity is excited by a single strong planet–planet scattering (Wu and Lithwick 2011). It should be noted, however, that the peak observed by Santerne et al. (2016) may be broader than expected from the high-eccentricity migration scenario, and could be accommodated in the disk migration scenario as well.

3.1.2.3 Hot Jupiters Are Not So Lonely

The "fact" that hot Jupiters are rarely accompanied by close siblings has often been cited as evidence for the high-eccentricity migration scenario, which requires the absence of close companions (e.g., Wu and Murray 2003) and/or clears away the close companions during the process. A recent study by Schlaufman and Winn (2016), however, showed that it is actually not the case. They computed the conditional probability that Jupiter-sized planets with various orbital periods have another planet in the same system and found that hot Jupiters are as likely as longer-period Jupiters to have companions inside the snow line. This argues against the high-eccentricity migration except for some of its variant including the scattering after the inward disk migration (Guillochon et al. 2011).

As for the Kozai migration, detection of a distant stellar companion provides an indirect support for the theory (e.g., Wu and Murray 2003). While the high observed tertiary rate of spectroscopic binaries with periods less than three days (Tokovinin et al. 2006) seems to support the KCTF as the formation scenario of shortest-period binaries, no significant correlation has been found so far between the companion rate and the occurrence of short-period giant planets exhibiting significant eccentricities

3.1 High-Eccentricity Migration 41

and/or spin–orbit misalignments (Knutson et al. 2014; Ngo et al. 2015; Piskorz et al. 2015). The fact may also indicate that the secular perturbation plays at most a supporting role in the formation of hot Jupiters.

3.1.2.4 Paucity of Super-Eccentric Warm Jupiters

If hot Jupiters are mainly formed through the high-eccentricity migration from beyond the snow line, there should also exist warm Jupiters on highly eccentric orbits that are undergoing tidal migration (Socrates et al. 2012). Dawson et al. (2015) showed that the expected population does not exist in the *Kepler* data based on the transit duration statistics. This argues against the high-eccentricity scenario, at least in its simplest form, as a dominant channel of hot-Jupiter formation. That said, some of its variant are not necessarily excluded. For example, high-eccentricity migration may have started interior to ~ 1 AU after disk migration (e.g., Guillochon et al. 2011); warm Jupiters' eccentricities may be currently undergoing secular oscillations due to close companions (Dong et al. 2014) and they may be "super-eccentric" only for a fraction of time; or tidal circularization might occur more rapidly at higher eccentricities, reducing the number of planets in highly-eccentric orbits.

3.1.2.5 Difficulty in Producing Counter-Orbiting Hot Jupiters

As shown in Fig. 1.5, some hot Jupiters have λ close to 180°. It has been shown that even the high-eccentricity migration is difficult to produce such "counter-orbiting" hot Jupiters with $\psi \approx 180°$ (Xue and Suto 2016). We should note, however, that λ is a sky-projection of the true obliquity ψ, and λ for the retrograde orbit with $\psi > 90°$ tends to be larger than the true obliquity ψ (Fabrycky and Winn 2009). In other words, the planets with $\lambda \approx 180°$ may actually have smaller ψ compatible with the high-eccentricity migration scenario. We will show that it is indeed the case for at least one of those systems, HAT-P-7, in Chap. 4.

3.2 Tidal Origin of the Obliquity Trend

Suppose that the high-eccentricity migration is responsible for the spin–orbit misalignments, can it also explain the observed obliquity trends, especially the λ–T_{eff} correlation (Sect. 2.4.1)? No convincing arguments have been presented that explain why the above mechanisms for the high-eccentricity migration could produce such a steep dependence of stellar obliquity on the host-star property. Instead, the obliquity trend may be attributed to the difference in the stellar property as follows.

Even after the orbit circularization, tidal evolution continues. Tides raised on the star by the planet, which are much weaker than the planetary counterpart, instead start to play a role. As was also the case for planetary tides, stellar tides synchronize the stellar rotation with the orbital motion, damp the spin–orbit misalignment if any, and lead to the decay of the semi-major axis.

Winn et al. (2010) proposed that the observed λ–T_{eff} relation may be explained by this tidal damping. The scenario is based on the fact that the mass of the convective envelope starts to drop significantly above $T_{\text{eff}} \sim 6100$ K (Pinsonneault et al. 2001). Since the turbulence in the convective layer is thought to greatly enhance the tidal dissipation efficiency (e.g., Zahn 2008), the obliquity damping is also pronounced around cooler stars. The magnetic field produced by the convective envelope, and the resulting braking of the stellar rotation, may also explain why the stars with low obliquities, if tidally aligned, are *not* synchronized with the planetary orbit but rotates more slowly.[5] Indeed, the same temperature also corresponds to the so-called "Kraft break," below which stellar rotation speed sharply decreases (Kraft 1967; Gray 2005).[6]

Qualitatively, the trends we have discussed in Sect. 2.4 are in agreement with the hypothesis. The correlation with the stellar temperature or age, along with the outliers with large values of $a/R_\star$ (Fig. 2.5), are consistent with the tidal damping. The lower obliquities observed for the most massive planets (Fig. 2.6) may also naturally arise from the stronger tides raised by more massive planets.

An attempt was made by Albrecht et al. (2012) to establish a single quantitative measure that explains these trends. Because the current understanding of the tidal dissipation limits the realistic computation of this timescale from the first principles, they adopted simple scaling laws for the tidal *synchronization* timescales by Zahn (1977) and showed that this timescale beautifully sorts the systems in order of the degree of their spin–orbit misalignments (their Fig. 24), although the absolute timescales are rather arbitrarily chosen.[7] The argument suggests that the tidal star–planet interaction plays an important role in sculpting the observed obliquity distribution, whether the *damping* is indeed responsible or not.

3.2.1 Possible Evidence Against the Tidal Origin

Here we discuss two main difficulties of the tidal scenario for the λ–T_{eff} trend. Neither of them is decisive, though, mainly due to the unknown nature of tides.

---

[5]The orbital periods of hot Jupiters are less than a week, while the rotation periods of their host stars are typically $\mathcal{O}(10)$ days.

[6]Dawson (2014) advocated that this rapid decrease in the stellar spin angular momentum, rather than the efficiency of tidal dissipation, is responsible for the trend.

[7]The computed timescales, in their original forms, are by many orders of magnitudes longer than the system age. This might be due to the difference in the timescales for spin–orbit synchronization and circularization.

3.2.1.1 Weak Dependence of Obliquity on the Orbital Distance Around Cool Stars

If the obliquity is indeed damped by the tidal interaction, planets closer to their host stars should exhibit better spin–orbit alignments. While this may be the case for the Rossiter-McLaughlin (RM) sample around cool stars (blue circles in Fig. 3.1), the analysis of spot-modulation amplitudes of *Kepler* stars (Mazeh et al. 2015, Sect. 2.3.5) did *not* find any significant difference between the obliquity distributions of planets with periods 1–5 days and those with 5–50 days. Although the more in-depth analysis based on the same technique (Li and Winn 2016) identified a statistically significant correlation with the orbital periods that is qualitatively consistent with the tidal damping (i.e., decreasing obliquity with the decreasing orbital period), the trend is still quantitatively inconsistent with the tidal scenario; the trend they found is rather smooth and extends up to the orbital period of ∼30 days, while the tidal scenario predicts a steep decrease in the obliquity at a much shorter period.

While these results may argue against the tidal realignment scenario, we should note that the property of the sample in the above inferences is not the same as the RM one. The former sample includes the whole KOIs and so most of the planets discussed here are much smaller than hot Jupiters (cf. Sect. 1.1). Thus, the tides raised on the star, which are supposed to be responsible for the tidal damping, are also smaller, and this weaker tide can be consistent with the weak signature of tidal interaction at least qualitatively. In any case, the correlation found by Li and Winn (2016) indicates that our understanding of the obliquity distribution still lacks some important process.

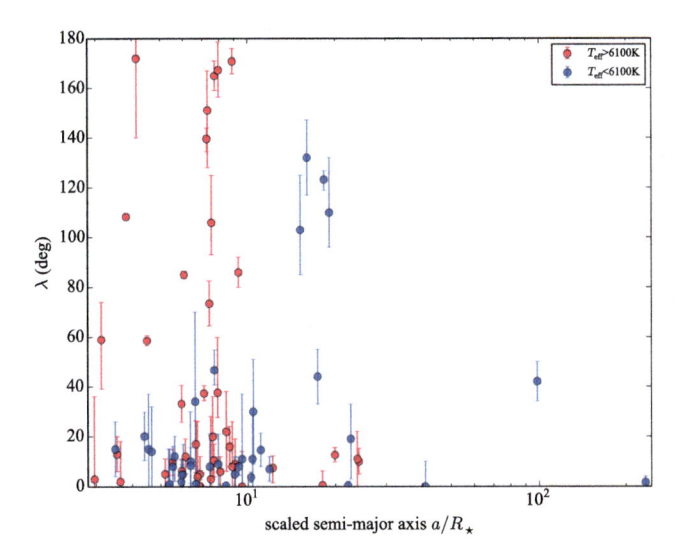

Fig. 3.1 The values of λ as a function of the semi-major axis divided by the stellar radius $a/R_\star$ (the same sample as in the previous figures). Blue circles correspond to planets around stars with $T_{\text{eff}} < 6100$ K, while red ones are those around stars with $T_{\text{eff}} > 6100$ K

3.2.1.2 Tidal Realignment Involves Tidal Orbital Decay

Winn et al. (2010) already pointed out that the planet must surrender such large angular momentum to realign the star that it would be engulfed by the star when the realignment is completed, at least according to the simplest tidal model. Lai (2012) proposed a solution to this problem by presenting a new tidal model, where the spin–orbit realignment occurs on a different timescale from the orbital decay. The evolution simulations based on this model (Rogers and Lin 2013; Xue et al. 2014), however, showed that even the revised model is inconsistent with the current observed distribution of λ, at least in its simple form. Li and Winn (2016) performed a parameter search and confirmed that it requires fine tuning for the spin–orbit realignment to occur earlier than the orbital decay. Currently it is not clear whether this inconsistency is due to the incompleteness of the tidal model or simply indicates that the tidal realignment scenario is wrong.

3.3 Star–Disk Misalignment

So far, we have discussed the scenarios that the spin–orbit misalignment is "acquired" due to the orbital evolution after the planet formation, implicitly assuming that the stellar spin axis is initially well aligned with the axis of the protoplanetary disk, and hence with the orbital axes of the planets formed in it. The initial star–disk alignment indeed seems to be the case for our solar system, and conforms well with a naive expectation from the simple theory of disk formation. In addition, good spin–orbit alignments in multi-transiting systems (Sect. 2.4.3) also seem to support the universality of the notion. Nevertheless, it would still be valuable to investigate alternatives, given the several pieces of evidence that possibly argue against the high-eccentricity migration (Sect. 3.1.2) and the subsequent tidal realignment (Sect. 3.2.1).

Indeed, several mechanisms have also been proposed to produce the "primordial" misalignment, that is, the misalignment between the axes of stellar spin and protoplanetary disk. If this really happens, the high-eccentricity migration is not necessarily required to explain the observed spin–orbit misalignment, and the formation of hot Jupiters may entirely be explained by the smooth disk migration and/or in-situ formation. In this section, we comment on this class of scenarios.

3.3.1 Possible Origins of Primordial Misalignment

3.3.1.1 Disk Torquing and Magnetic/Gravitational Star–Disk Interaction

If a disk-hosting star has a companion star whose orbit is inclined by angle I with respect to the initial protoplanetary-disk plane, the perturber's gravity may torque the

3.3 Star–Disk Misalignment

disk out of the stellar equatorial plane (Batygin 2012). Specifically, both stellar spin and protoplanetary disk axes precess around the axis of the outer binary, whose orbit dominates the angular momentum budget of the whole system, and the difference in the precession rates periodically induces the spin–orbit misalignment by $2I$ at maximum.

The subsequent study by Batygin and Adams (2013) considered the combined effect of disk torquing, gravitational disk–star coupling due to the quadrupole moment of the rapidly-rotating pre main-sequence (PMS) star, disk accretion, and the rotational evolution of the central PMS star due to gravitational contraction as well as magnetic braking. They found that resonance between the precession frequencies of disk-torquing and spin-precession makes it possible to excite the spin–orbit misalignment even from an initially small value, unlike the previous case where the torquing from a companion alone is considered.

Furthermore, Spadling and Batygin (2014) and Lai (2014) independently showed that the torque due to the magnetosphere–disk interaction (Lai et al. 2011), if taken into account in the above scheme, leads to even more diverse spin-axis evolution. Spalding and Batygin (2015) proposed that the scenario possibly explains the λ–T_{eff} trend rather as the correlation with the stellar mass (Fig. 3.2). They pointed out that the magnetic torques act to realign the stellar spin axis, and that massive T-Tauri stars tend to have weaker magnetic dipole fields than their less-massive counterparts (Gregory et al. 2012); hence the primordial misalignment is preserved for massive stars while it is washed out for low-mass stars.

As was the case for the Kozai migration, this scenario also predicts the correlation between the companion occurrence and spin–orbit misalignment, which appears to contradict the observation (see Sect. 3.1.2). It is however possible that the companion responsible for the primordial misalignment is lost before the planet formation, due to the complex dynamics of stellar clusters that may sometimes lead to dissolution of multi-stellar systems (Spalding and Batygin 2014). Additional computational efforts are required to assess the validity of this explanation.

3.3.1.2 Chaotic Accretion

Due to the turbulence in the star-forming environment, the angular momentum of the protoplanetary disk, which is usually dominated by the last-accreted gas, may have different direction from that of the star, which is determined by the sum of the accreted angular momentum (Bate et al. 2010; Fielding et al. 2015). These simulations, however, do not take into account the star–disk interaction properly, because of very different time scales to solve the stellar structure and the disk. Indeed, the semi-analytic model incorporating this aspect shows that the protoplanetary disk, though occasionally tilted away from the stellar equator due to turbulence, will eventually be aligned with the stellar equator when the accretion ceases (Spalding et al. 2014), questioning the viability of the mechanism.

3.3.1.3 Internal Gravity Waves

Rogers et al. (2012) proposed[8] that the angular momentum transport due to the internal gravity waves (IGWs) modulates the surface rotation of the star to produce apparent spin–orbit misalignments. The mechanism works preferentially for hot stars because the IGW is generated at the boundary of convective cores and radiative envelopes of hot stars. This scenario, however, is not supported observationally because the radial differential rotation as predicted by this process has not been observed for main-sequence stars exhibiting spin–orbit misalignments (Benomar et al. 2014, 2015).

3.3.2 Obliquity Trends in the Primordial Misalignment Scenario

Whether the spin–orbit misalignment is primordial or not, the tidal scenario can be invoked to explain the observed obliquity trend. On the other hand, the disk-torquing and IGW scenarios contain the internal mechanisms that produce the correlation. These scenarios, if more thoroughly investigated, may well be appealing enough given the difficulties in the current tidal scenario and fewer assumptions required. Here let us make brief comments on the difference in interpreting the obliquity trend as a part of the primordial misalignment scenario, rather than the tidal realignment scenario in Sect. 3.2.

First, both disk torquing and IGW scenarios explain the correlation as that with stellar mass, rather than with effective temperature. This does not affect the overall feature of the trend, as the two are well correlated for the main-sequence stars (Fig. 3.2).

On the other hand, interpretation of the known "exceptions", or the prediction for the properties of exceptional cases, is different. In the tidal realignment scenario, planets with large $a/R_\star$ or small mass are allowed to be exceptions to the trend because of the long timescale for the tidal damping. In the primordial scenarios, the four clear exceptions need to be explained in a different manner. Spalding and Batygin (2015) argued that the eccentricities of their orbits are the imprint of their past dynamical interactions, and some spin–orbit misalignments would be due to the dynamical origin, rather than primordial; this argument may be supported by Fig. 3.2, in which we distinguish the planets with non-zero observed eccentricities by open circles.[9] Spalding and Batygin (2015) thus predict that the planets on circular orbits should basically follow the λ–$M_\star$ trend, while eccentric ones do not need to be the case. As shown in Fig. 3.3, the correlation between λ and eccentricity is currently unclear, and future observations will confirm or reject this hypothesis.

---

[8]This mechanism is not exactly to produce the star–disk misalignment, but we discuss it here because it does *not* alter the planetary orbit but tilt the stellar spin with respect to the *initial* disk plane.
[9]Kepler-63 has only an upper limit $e < 0.43$ for the eccentricity.

3.3 Star–Disk Misalignment

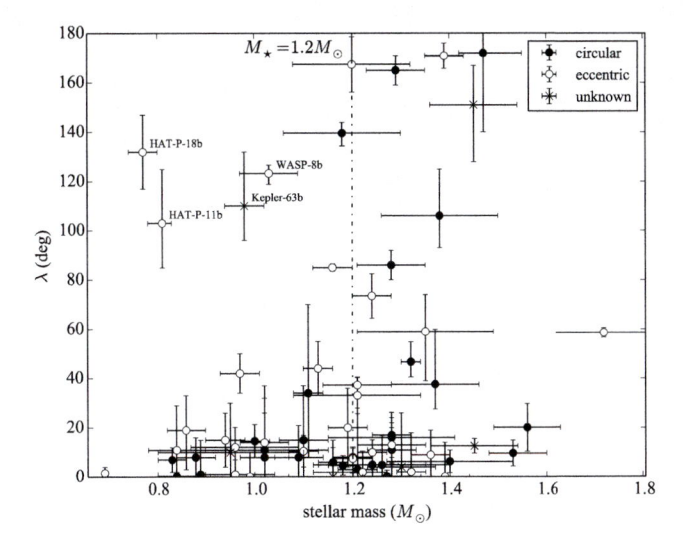

Fig. 3.2 The values of λ as a function of the host-star mass. For planets drawn with open circles, non-zero eccentricities have been detected at more than 1σ level. Planets shown with crosses have non-zero upper limits on their eccentricities

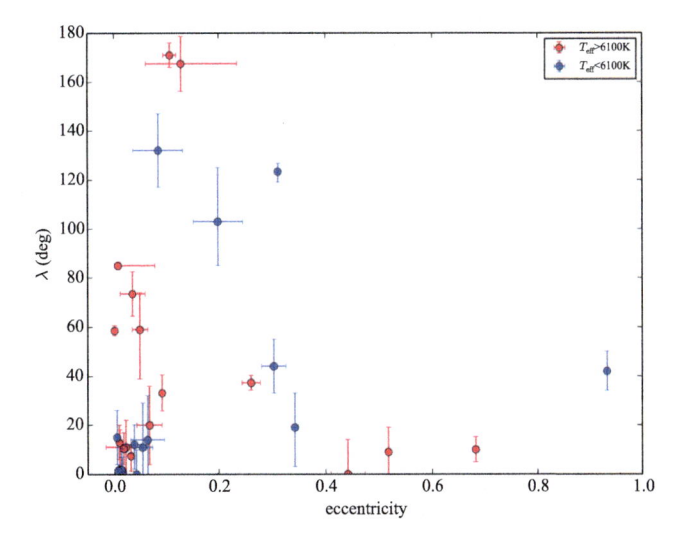

Fig. 3.3 The values of λ as a function of the orbital eccentricity. Blue and red colors show that their host stars are cooler and hotter than 6100 K, respectively

Finally, the primordial scenarios have no mechanisms to produce the planetary-mass cut off for the retrograde orbits, at least in their current forms. Additional assumption, whether it is tidal interaction or different formation, therefore seems necessary if this cut off is real.

3.4 Summary and Outstanding Questions

In this chapter, we have discussed the possible explanations for the spin–orbit misalignment of hot Jupiters and for the observed correlation(s) between the misalignment and stellar properties. The discussions so far are summarized as follows:

- High-eccentricity migration, the excitation of a high eccentricity due to few-body dynamical processes followed by the tidal orbit circularization, can explain the existence of both hot Jupiters and eccentric planets.
- While it may not be a dominant mechanism to produce hot Jupiters given the lack of observational supports, the scenario provides the most natural explanation for the current architectures of at least a few systems. It is quantitatively unclear to what extent the observed misalignments could be due to the high-eccentricity migration.
- Tidal star–planet interactions seem to explain the known obliquity trends at least qualitatively. The scenario is, however, still incomplete quantitatively, mainly due to the uncertain nature of the tidal interaction.
- Given the situation, the scenarios that the spin–orbit misalignment is the remnant of the primordial star–disk misalignment remain to be viable alternatives.

Below we address several questions for future studies that will be of importance for solving the mysteries.

3.4.1 Are Hot Jupiters Special?

If the spin–orbit misalignment is indeed linked to the high-eccentricity migration, it should be a property specific to hot Jupiters. If the misalignment is primordial, on the other hand, it should be observed for any sort of planetary systems, not limited to hot Jupiters.[10] To distinguish the two scenarios, therefore, it is crucial to measure obliquities for stars hosting planets other than hot Jupiters.

As mentioned in Chap. 2, most of the current obliquity measurements for individual systems, which are mostly from the RM measurement, are for hot Jupiters. Some of the 15 RM measurements made for planets with orbital periods longer than 7 days or with masses less than $0.3\,M_{\text{Jup}}$ exhibit spin–orbit misalignments (Fig. 3.4). It is, therefore, not yet clear whether the high obliquity is indeed specific to hot Jupiters. Especially, there are only few measurements for longer-period planets around hot stars, for which most of the misaligned systems have been observed (Fig. 3.5). If large obliquities are not common in this area, it becomes unlikely that the misalignment is primordial, because there is no reason in the primordial scenario that only the close-in planets are preferentially misaligned.

---

[10] If the latter is the case, the low stellar obliquity in our solar system turns out to be a coincidence, rather than the norm.

3.4 Summary and Outstanding Questions   49

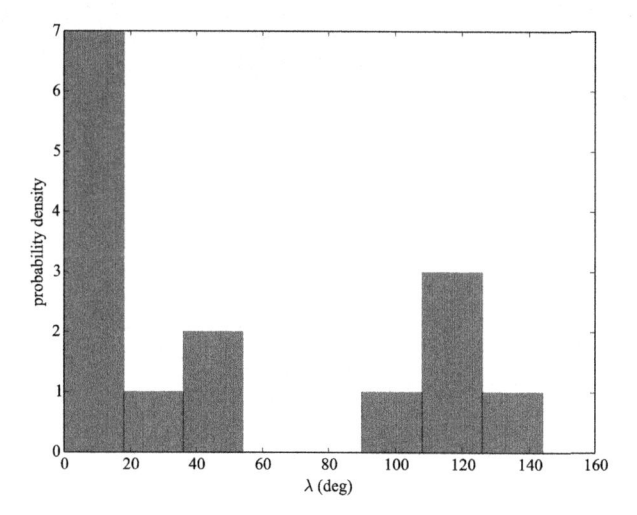

Fig. 3.4 Histogram of λ for a subset of the RM sample in Fig. 1.5 consisting of planets with orbital periods longer than 7 days or with masses less than 0.3 M_{Jup}

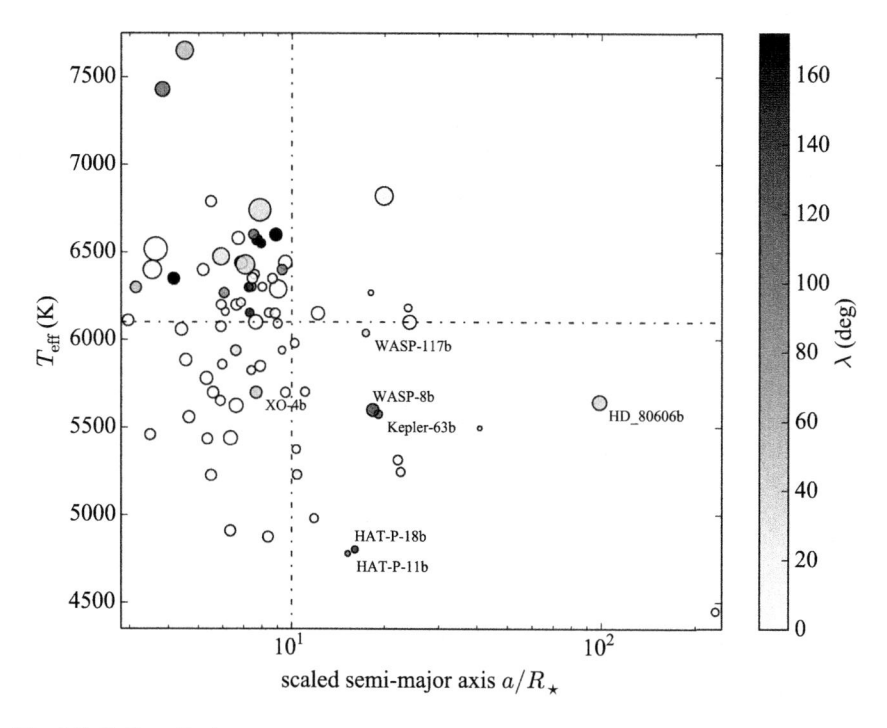

Fig. 3.5 Stellar effective temperature T_{eff} versus scaled semi-major axis $a/R_\star$ for systems with λ measurements. The color of each circle corresponds to the value of λ, and its area is proportional to the planetary mass

Extending the RM observations to longer-period planets is difficult for two practical reasons. First, a long-period planet rarely transits, and even if it does, it is not guaranteed whether the transit can be observed from a suitable ground-based facility. Even if the transit is observable, it may not be enough; the second difficulty comes from the long transit duration. While the whole shape of the in-transit RV anomaly needs to be captured for a reliable measurement of λ, transit durations of planets with period longer than 10 days are often comparable to the length of one night (cf. Eq. 1.6). If the duration is too long, therefore, we may need to observe the transit multiple times.

For this reason, it will be of great advantage if the transit data from the *Kepler* space telescope can be utilized for the stellar obliquity measurements of longer-period planets. The low transit probability and rareness of transits are compensated by the long-term, continuous observations of a large number of stars, and the long transit duration does not matter at all for the space-based observations. In this thesis, such methodologies will be discussed in Chaps. 4 and 5.

3.4.1.1 Stellar Obliquities in Multi-Transiting Systems

The most crucial test for the primordial misalignment scenario would be the stellar obliquity measurements for multi-transiting systems. Since the orbital planes of multiple transiting planets are likely well aligned a priori, these planes most likely trace the original protoplanetary disks. Thus, if a high obliquity is found for any one of the planets in multi-transiting systems, that can be evidence for the primordial star–disk misalignment (Lai et al. 2011; Albrecht et al. 2012). As mentioned in Sect. 2.4, such tests are already underway, and multi-transiting systems so far exhibit low obliquities. Nevertheless, we should note that the number of systems is still small (with one out of seven systems exhibiting a clear misalignment) and that the alignments for the other six systems are based on the two-dimensional (i.e., sky-projection or line-of-sight component alone) measurements; the latter problem is revisited for Kepler-25 in Chap. 4.

There still remain two important questions regarding the exception, Kepler-56, which is the only multi-transiting system with strong evidence of a spin–orbit misalignment. First, what is the origin of the misalignment in this system? The simplest answer is the initial star–disk misalignment as discussed in Sect. 3.3. However, it has also been proposed that an outer companion detected in the long-term RV trend of Kepler-56 could have torqued the planetary orbits out of the stellar equatorial plane, if the companion's orbit is misaligned with the stellar equator (Huber et al. 2013). A more detailed investigation of the dynamical scenario that could produce the spin–orbit misalignment in a multi-transiting system is thus crucial to find the "smoking gun" for the star–disk misalignment.

Second, are systems like Kepler-56 indeed rare? To answer this question, we note that the host stars of the other six "aligned" multi-transiting systems basically have T_{eff} around or below the threshold, 6100 K, below which spin–orbit alignments are the norm (Sect. 2.4.1). In this regard, it is suggestive that the only exception Kepler-56 is a slightly massive evolved star with $M_\star = 1.32 \pm 0.13\, M_\odot$ (Huber et al. 2013), which would have belonged to the population of "hot" stars in its main sequence phase. Thus it still seems possible that multi-transiting systems with significant spin–orbit misalignments more frequently exist around hotter stars, as in the overall trend, and that such systems currently look rare simply because we are focusing on cool stars, as was the case in the history of the RM measurements (Winn et al. 2010). If such a trend is confirmed in future, it will be strong evidence that supports the primordial origin of the spin–orbit misalignment, as described in Sect. 3.3.

3.4.2 Are All Planetary Systems Flat?

So far, no multi-planetary system has been shown to have a significant mutual orbital inclination with high confidence. This is partly due to its construction, because most of the known multi-planetary systems are multi-*transiting* systems discovered by *Kepler*, who are unlikely to be observed as such if their orbits were mutually inclined.

In contrast, the mechanisms involved in high-eccentricity migration scenario (i.e., planet–planet scattering, planetary Kozai effect, and secular chaos), if indeed at work, should produce multi-planetary systems with two planets on mutually inclined (and possibly eccentric and widely separated) orbits. If the innermost planet obtains an eccentricity close to unity, its orbit is circularized to become a close-in planet, as described in Sect. 3.1, and decoupled from the outer planet. This also implies that the innermost planet, if fails to obtain a sufficiently large eccentricity, may not become a close-in planet but retain a modest hierarchy with the outer planet (e.g., Dong et al. 2014).

While indirect arguments suggest such "orbit–orbit misalignments" for several systems (Dawson and Chiang 2014) with the modest hierarchy, no direct detection has been presented so far. Although it is generally difficult to constrain the mutual orbital inclination, the *Kepler* data continuously obtained for four years may reveal such systems through the transit variations due to the gravitational interaction, as has been made possible for hierarchical triple-star systems (e.g., Borkovits et al. 2016, see also Chap. 6). Such an architecture, if detected, could be an important clue to understand the dynamical evolution of planetary systems.

3.4.3 Initial Distribution of the Star–Disk Misalignment

Even without the perturbing companion, the star–disk misalignment may be acquired during the disk formation, as mentioned in Sect. 3.3.1. Eventually, the initial distribution will be needed to give a complete answer to this nature and nurture problem. It is currently controversial whether the star–disk misalignment is the norm or not, mainly due to the difficulty in solving the disk formation and star–disk interaction (rotational evolution of the star) in a consistent manner. Such a simulation is still computationally too heavy, and some breakthrough might be required for a realistic one.

Another possible approach is to directly observe the star–disk misalignment. Inclination measurements for 18 *debris* disks using their resolved images suggest that the star–disk alignment is the norm (Watson et al. 2011; Greaves et al. 2014), on the basis of a presumably natural assumption that debris disks share the same plane with protoplanetary disks. Even the planetesimal belts are resolved by recent observations of HR 8799 with ALMA (Booth et al. 2016), in which orbits of four directly images planets out to $\sim$ 100 AU (Konopacky et al. 2016), stellar equator (Wright et al. 2011), and the disk are all co-aligned.

3.4.4 Efficiency of Tides

If the high obliquity is found to be specific to hot stars with hot Jupiters, it becomes more likely to be related to their migration process. The next question is the origin of the temperature dependence. Tidal realignment discussed in Sect. 3.2 is one possibility, but its plausibility is unclear mainly because efficiency of the tidal damping is theoretically quite uncertain: even the timescale for tidal dissipation is not understood from first principles. Any observational constraint on the tidal dissipation, or the statistical inference based on the properties of existing systems, is therefore of great importance.

One possible approach is to monitor the orbital period of a close-in planet with small $a/R_\star$ to detect the decrease in the orbital period due to tidal dissipation. Indeed, if the efficiency of tides is close to the currently expected value or larger, such an "orbital decay" could be observable over the timescales of a few years for the most favorable targets, as recently claimed by Maciejewski et al. (2016).

In fact, if the tidal orbital decay and the resulting ingestion of the close-in planet is common, that may be another explanation for the obliquity trend. Matsakos and Königl (2015) pointed out that the angular momentum surrendered to the star by an ingested planet is enough to realign the cool central star, which rotates rather slowly and has smaller angular momentum, while hotter stars would hardly be affected by the engulfed planet. This scenario is similar to the tidal realignment scenario discussed in Sect. 3.2, although the realignment is caused by the engulfment rather than dissipation. An advantage of this scenario is that it is consistent with the universality of the λ–T_{eff} correlation and the lack of strong period dependence of obliquity around cool stars discussed in Sect. 2.4. This is because the degree of realignment does not have to be related to the property of currently existing planets that survived the engulfment. That feature could also be a weakness because we do see some correlations between λ and the properties of current planetary systems, though they are in general less clear than λ–T_{eff} trend. This scenario, as is the case for the tidal realignment, also relies on the picture that spin–orbit misalignments are initially universal; both high-eccentricity migration or initial star–disk misalignment thus qualify.

References

S. Albrecht, J.N. Winn, J.A. Johnson et al., ApJ **757**, 18 (2012)
M.R. Bate, G. Lodato, J.E. Pringle, MNRAS **401**, 1505 (2010)
K. Batygin, Nature **491**, 418 (2012)
K. Batygin, F.C. Adams, ApJ **778**, 169 (2013)
O. Benomar, K. Masuda, H. Shibahashi, Y. Suto, PASJ **66**, 94 (2014)
O. Benomar, M. Takata, H. Shibahashi, T. Ceillier, R.A. García, MNRAS **452**, 2654 (2015)
M. Booth, A. Jordán, S. Casassus et al., MNRAS, arXiv:1603.04853 (2016)
T. Borkovits, T. Hajdu, J. Sztakovics et al., MNRAS **455**, 4136 (2016)
S. Chatterjee, E.B. Ford, S. Matsumura, F.A. Rasio, ApJ **686**, 580 (2008)
A.C.M. Correia, ApJ **704**, L1 (2009)
A. Cumming, G.W. Marcy, R.P. Butler, ApJ **526**, 890 (1999)
R.I. Dawson, ApJ **790**, L31 (2014)
R.I. Dawson, E. Chiang, Science **346**, 212 (2014)
R.I. Dawson, R.A. Murray-Clay, ApJ **767**, L24 (2013)
R.I. Dawson, R.A. Murray-Clay, J.A. Johnson, ApJ **798**, 66 (2015)
S. Dong, B. Katz, A. Socrates, ApJ **781**, L5 (2014)
P.P. Eggleton, L. Kiseleva-Eggleton, ApJ **562**, 1012 (2001)
D. Fabrycky, S. Tremaine, ApJ **669**, 1298 (2007)
D.C. Fabrycky, J.N. Winn, ApJ **696**, 1230 (2009)
D.B. Fielding, C.F. McKee, A. Socrates, A.J. Cunningham, R.I. Klein, MNRAS **450**, 3306 (2015)
E.B. Ford, B. Kozinsky, F.A. Rasio, ApJ **535**, 385 (2000)
D.F. Gray, The Observation and Analysis of Stellar Photospheres (2005)
J.S. Greaves, G.M. Kennedy, N. Thureau et al., MNRAS **438**, L31 (2014)
S.G. Gregory, J.-F. Donati, J. Morin et al., ApJ **755**, 97 (2012)
J. Guillochon, E. Ramirez-Ruiz, D. Lin, ApJ **732**, 74 (2011)
M. Holman, J. Touma, S. Tremaine, Nature **386**, 254 (1997)
A.W. Howard, G.W. Marcy, S.T. Bryson et al., ApJS **201**, 15 (2012)
D. Huber, J.A. Carter, M. Barbieri et al., Science **342**, 331 (2013)
M. Jurić, S. Tremaine, ApJ **686**, 603 (2008)
L.G. Kiseleva, P.P. Eggleton, S. Mikkola, MNRAS **300**, 292 (1998)
H.A. Knutson, B.J. Fulton, B.T. Montet et al., ApJ **785**, 126 (2014)
Q.M. Konopacky, C. Marois, B.A. Macintosh et al., arXiv e-prints arXiv:1604.08157 (2016)
Y. Kozai, AJ **67**, 591 (1962)
R.P. Kraft, ApJ **150**, 551 (1967)
D. Lai, MNRAS **423**, 486 (2012)
D. Lai, MNRAS **440**, 3532 (2014)
D. Lai, F. Foucart, D.N.C. Lin, MNRAS **412**, 2790 (2011)
G. Li, S. Naoz, B. Kocsis, A. Loeb, ApJ **785**, 116 (2014)
G. Li, J.N. Winn, ApJ **818**, 5 (2016)
D.N.C. Lin, P. Bodenheimer, D.C. Richardson, Nature **380**, 606 (1996)
G. Maciejewski, D. Dimitrov, M. Fernández et al., A&A **588**, L6 (2016)
F. Marzari, S.J. Weidenschilling, Icarus **156**, 570 (2002)
T. Matsakos, A. Königl, ApJ **809**, L20 (2015)
T. Mazeh, H.B. Perets, A. McQuillan, E.S. Goldstein, ApJ **801**, 3 (2015)
T. Mazeh, J. Shaham, A&A **77**, 145 (1979)
C.D. Murray, S.F. Dermott, *Solar system dynamics* (Cambridge University Press, 1999)
M. Nagasawa, S. Ida, T. Bessho, ApJ **678**, 498 (2008)
S. Naoz, W.M. Farr, Y. Lithwick, F.A. Rasio, J. Teyssandier, Nature **473**, 187 (2011)
H. Ngo, H.A. Knutson, S. Hinkley et al., ApJ **800**, 138 (2015)
C. Petrovich, ApJ **805**, 75 (2015)
M.H. Pinsonneault, D.L. DePoy, M. Coffee, ApJ **556**, L59 (2001)

D. Piskorz, H.A. Knutson, H. Ngo et al., ApJ **814**, 148 (2015)
F.A. Rasio, E.B. Ford, Science **274**, 954 (1996)
T.M. Rogers, D.N.C. Lin, ApJ **769**, L10 (2013)
T.M. Rogers, D.N.C. Lin, H.H.B. Lau, ApJ **758**, L6 (2012)
A. Santerne, C. Moutou, M. Tsantaki et al., A&A **587**, A64 (2016)
K.C. Schlaufman, J.N. Winn, ArXiv e-prints arXiv:1604.03107 (2016)
A. Socrates, B. Katz, S. Dong, S. Tremaine, ApJ **750**, 106 (2012)
C. Spalding, K. Batygin, ApJ **790**, 42 (2014)
C. Spalding, K. Batygin, ApJ **811**, 82 (2015)
C. Spalding, K. Batygin, F.C. Adams, ApJ **797**, L29 (2014)
N.I. Storch, K.R. Anderson, D. Lai, Science **345**, 1317 (2014)
G. Takeda, F.A. Rasio, ApJ **627**, 1001 (2005)
A. Tokovinin, S. Thomas, M. Sterzik, S. Udry, A&A **450**, 681 (2006)
S. Udry, M. Mayor, N.C. Santos, A&A **407**, 369 (2003)
C.A. Watson, S.P. Littlefair, C. Diamond et al., MNRAS **413**, L71 (2011)
S.J. Weidenschilling, F. Marzari, Nature **384**, 619 (1996)
J.N. Winn, D. Fabrycky, S. Albrecht, J.A. Johnson, ApJ **718**, L145 (2010)
D.J. Wright, A.-N. Chené, P. De Cat et al., ApJ **728**, L20 (2011)
Y. Wu, Y. Lithwick, ApJ **735**, 109 (2011)
Y. Wu, N. Murray, ApJ **589**, 605 (2003)
Y. Wu, N.W. Murray, J.M. Ramsahai, ApJ **670**, 820 (2007)
Y. Xue, Y. Suto, ApJ **820**, 55 (2016)
Y. Xue, Y. Suto, A. Taruya et al., ApJ **784**, 66 (2014)
J.-P. Zahn, A&A **57**, 383 (1977)
J.-P. Zahn, in EAS Publications Series, vol. 29, ed. M.-J. Goupil, J.-P. Zahn, pp. 67–90

Chapter 4
Three-Dimensional Stellar Obliquities of HAT-P-7 and Kepler-25 from Joint Analysis of Asteroseismology, Transit Light Curve, and the Rossiter–McLaughlin Effect

Abstract Measurements of stellar obliquities for transiting systems are usually two-dimensional: either the sky-projection λ of the true obliquity, or the difference between orbital inclination (almost $90°$) and stellar inclination $i_\star$, is used to infer the degree of the spin–orbit misalignment. In this chapter, we develop a methodology for determining true stellar obliquity ψ, combining the analyses of asteroseismology, transit light curves, and the Rossiter–McLaughlin effect. We demonstrate the power of such a joint analysis by applying it for the first time to two real systems, HAT-P-7 hosting a hot Jupiter and Kepler-25 with two transiting planets and another non-transiting one. We also show that the joint analysis allows for an accurate and precise determination of the numerous parameters characterizing the planetary system, in addition to ψ.

Keywords Asteroseismology · The Rossiter–McLaughlin effect · Occultation
HAT-P-7 · Kepler-25

4.1 Introduction

4.1.1 A Historical View on Measurements of λ

While the measurement of ψ is not easy, its projection onto the plane of the sky, λ, has already been measured for about 80 transiting planetary systems via the Rossiter–McLaughlin (RM) effect (Winn 2011, see also Sect. 2.2), and is now established as one of the most basic parameters that characterize transiting planetary systems; see Fig. 1.5 for the summary of current observations.

The RM effect was originally proposed to determine the projected spin–orbit angle of eclipsing binary star systems (Rossiter 1924; McLaughlin 1924). Queloz et al. (2000) successfully applied the technique for the first discovered transiting exoplanetary system, HD 209458, and obtained $\lambda = \pm 3°\!.9^{+18°}_{-21°}$. In the quest for improving the precision and accuracy, Ohta et al. (2005) presented an analytic formula to describe the RM effect and studied in detail the error budget and possible degeneracy among different parameters. This allowed Winn et al. (2005) to revisit HD 209458 with updated photometric and spectroscopic data, and to obtain $\lambda = -4°\!.4 \pm 1°\!.4$, improving the precision of the previous measurement by an order of magnitude.

In doing so, Winn et al. (2005) pointed out that the analytic approximation adopted by Ohta et al. (2005) leads to typically 10 percent error in the predicted velocity anomaly amplitude, while the estimated λ is fairly reliable. This motivated Hirano et al. (2010; 2011) to take into account stellar rotation, macroturbulence, and thermal/pressure/instrumental broadenings in modeling the stellar absorption line profiles. Those authors derived an analytic formula for the velocity anomaly of the RM effect by maximizing the cross-correlation function between the in-transit spectrum and the stellar template spectrum, i.e., following the same procedure as is actually used to derive RVs from the spectra. Their analytic formulae reproduce mock simulations within ∼0.5 percent, enabling the accurate and efficient multi-dimensional fit of parameters characterizing the star and planet(s) of an individual system.

More importantly, Winn et al. (2005) clearly demonstrated the potential of the RM effect to put strong quantitative constraints on the existing and/or future planetary formation scenarios. Indeed, when HD 209458 was the only known transiting planetary system, Ohta et al. (2005) discussed that *"Although unlikely, we may even speculate that a future RM observation may discover an extrasolar planetary system in which the stellar spin and the planetary orbital axes are anti-parallel or orthogonal. Then it would have a great impact on the planetary formation scenario, ..."*. In reality, however, they were too conservative. Among the 80 transiting planetary systems where the RM effect is observed, more than 30 exhibit significant spin–orbit misalignments with $|\lambda| > 22°\!.5$ (see Fig. 1.5). This unexpected diversity of the spin–orbit angle is not yet properly understood by the existing theories and remains an interesting challenge, as discussed in Chap. 3.

4.1.2 Aim: Determination of ψ

The main purpose of this chapter is to establish a methodology to determine ψ, instead of λ, through the joint analysis of asteroseismology, transit light curve, and the RM effect. We also present specific results for two interesting transiting planetary systems, HAT-P-7[1] (KIC 10666592) and Kepler-25 (KIC 4349452). HAT-P-7 is the first example of a system hosting a retrograde or a polar-orbit planet, while Kepler-25 is a multi-transiting system with three planets. We show that joint analyses of asteroseismology, transit light curve, and the RM effect provide stringent orbital parameter estimates as well as true stellar obliquity ψ.

As we noted in Sect. 2.1 and Fig. 2.1, λ differs from the true stellar obliquity ψ due to the projection onto the sky. Remember that, in addition to λ, ψ also depends

---

[1] We would like to emphasize the efforts made by Lund M. N. and his collaborators for their work on HAT-P-7. This system turned out to be studied simultaneously and independently by our respective teams.

4.1 Introduction

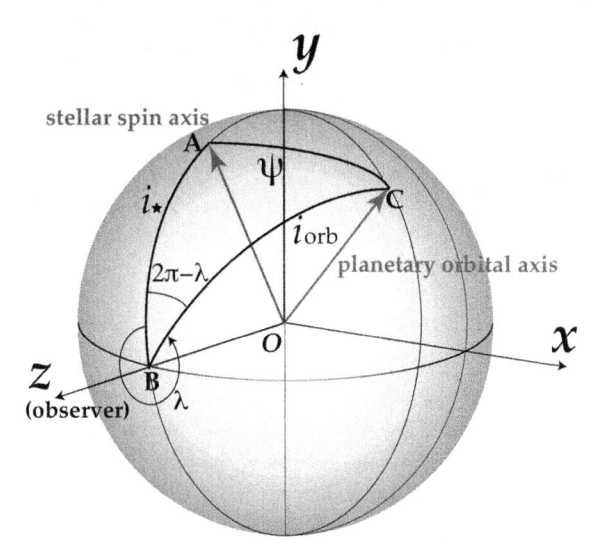

Fig. 4.1 Schematic illustration of geometric configuration of a star–planet system. We choose a coordinate system centered on the star, where the XY-plane is in the plane of the sky and $+Z$-axis points towards the observer. The $+Y$-axis is chosen along the sky-projected stellar spin and the X-axis is perpendicular to both Y- and Z-axes, forming a right-handed triad. Red and green arrows indicate, on a unit sphere, the angular momentum vectors of the stellar spin and the planetary orbital motion, respectively. The stellar and orbital inclinations, $i_\star$ and $i_{\rm orb}$, are measured from the $+Z$-axis and in the range of $[\,0°, 180°]$. The planetary orbital axis projected onto the sky plane is specified by the projected spin–orbit angle, λ, which is measured from the $+Y$-axis and in the range of $[0°, 360°]$. Note that λ is measured in the direction specified by the arrow. The angle AOC between the stellar spin and the planetary orbit axis vectors, ψ, is derived from the law of cosines for the spherical triangles ABC, as given by Eq. (4.1)

on the orbital inclination $i_{\rm orb}$ and the obliquity of the stellar spin axis $i_\star$. These angles are related by the law of cosines in spherical trigonometry,

$$\cos\psi = \cos i_\star \cos i_{\rm orb} + \sin i_\star \sin i_{\rm orb} \cos\lambda, \qquad (4.1)$$

as best illustrated in Fig. 4.1. In the case of transiting planetary systems, $i_{\rm orb}$ can be estimated from the transit light curve, and in any case is close to 90°. Given the projected angle λ measured from the RM effect, the major uncertainty for ψ therefore comes from the unknown stellar inclination $i_\star$. There are several complementary approaches to estimate $i_\star$, and hence ψ, as we already described in Sect. 2.3. In this chapter, we focus on asteroseismology (Unno et al. 1989; Aerts et al. 2010, see also Sect. 2.3.1)

In fact, target stars for the exoplanet hunting are often good targets for asteroseismology as well. In both transit and radial velocity surveys, low-mass, cool stars in the main sequence are usually favored, because their small radii are advantageous for the transit detection and because they have sharp and narrow absorption lines essential for the precise velocimetry. As is the case for the sun, such a low-mass, cool star has a thick convective envelope that sustains pulsations; turbulent motion as fast as the sound speed near the stellar surface stochastically generates acoustic waves, which propagate inside the star until they are damped. The oscillations with frequencies close to those of eigenmodes of the star are eventually sustained as many acoustic modes. Therefore, host stars with $T_{\text{eff}} \lesssim 7000$ K should commonly exhibit solar-like oscillations and allow for the application of asteroseismology.

4.1.3 Plan of This Chapter

This chapter is organized as follows. Section 4.2 summarizes the previous RM measurements and radial velocity (RV) data of the two systems. Section 4.3 presents a brief description on the procedure and results of the asteroseismology analysis, the latter of which will be used in the following joint analyses. Sections 4.4 and 4.5 analyze the *Kepler* transit light curves and the RV anomaly of the RM effect, using the asteroseismology results as the prior information, and show how the joint analysis improves the estimates of the system parameters. Section 4.6 is devoted to the summary and further discussion, and Sect. 4.7 concludes the chapter.

4.2 Previous Measurements of Stellar Obliquities

4.2.1 HAT-P-7

The HAT-P-7 system comprises a bright ($V = 10.5$) F6 star and a hot Jupiter transiting the host star with a 2.2-day period (hereafter P08 Pál et al. 2008). In addition to the significant spin–orbit misalignment first revealed by the Subaru spectroscopy (Narita et al. 2009; Winn et al. 2009), the fact that the system is in the *Kepler* field makes it very attractive as an asteroseismology target.

Interestingly, there have been three independent measurements of the RM effect for the HAT-P-7 system, which all indicate the significant spin–orbit misalignment, but do not agree quantitatively. Winn et al. (2009) (hereafter W09) performed the joint analysis of the spectroscopic and photometric transit of HAT-P-7b to obtain $\lambda = 182°.5 \pm 9°.4$. For RVs, they analyzed 17 spectra observed with the High Resolution Spectrograph (HIRES) on the Keck I telescope as well as 69 spectra observed with the High Dispersion Spectrograph (HDS) on the Subaru telescope. Eight of the HIRES spectra were from P08 and taken in 2007, while the other nine were obtained in 2009. Among 69 HDS spectra, 40 were obtained on 2009 July 1 that spanned a transit.

On the other hand, Narita et al. (2009) (hereafter N09) determined $\lambda = 227°.4^{+10°.5}_{-16°.3}$ (equivalently $\lambda = -132°.6^{+10°.5}_{-16°.3}$) based on the eight HIRES RVs from P08 and 40 HDS spectra spanning the transit on 2008 May 30. Although they fixed the transit parameters in the analysis of the RM effect, the systematics from the uncertainties of these parameters do not seem to explain the mild discrepancy with the W09 result, according to their discussion (see cases 1 to 4 in Sect. 4 of N09).

Later on, Albrecht et al. (2012) (hereafter A12) reported another measurement of the RM effect, resulting in $\lambda = 155° \pm 14°$. They analyzed 49 HIRES spectra spanning a transit on the night 2010 July 23/24 with the priors on transit parameters and ephemeris from the *Kepler* light curves.

In this chapter, we use the same RV data published in each of the three papers. Since the origin of the possible discrepancy in λ is not clear, we analyze each data set separately instead of combining the three.

4.2.2 Kepler-25

The Kepler-25 system is one of the few multi-transiting planetary systems with constrained λ. It consists of a relatively bright ($K_p = 10.7$) host star, two short-period Neptune-sized planets confirmed with transit timing variations (TTVs) (Steffen et al. 2012), and one outer non-transiting planet detected in a long-term RV trend (Marcy et al. 2014). Albrecht et al. (2013) (hereafter A13) measured $\lambda = 7° \pm 8°$ for the larger transiting planet Kepler-25c based on the HIRES spectra observed for two nights (2011 July 18/19 and 2012 May 31/June 1). Since the signal-to-noise ratio of the RV anomaly was small due to the relatively small radius of Kepler-25c, they also analyzed the time-dependent distortion of the spectral lines directly (i.e., Doppler tomography method in Sect. 2.2.2) and obtained a consistent result, $\lambda = -0°.5 \pm 5°.7$.

In this chapter, we analyze the RVs around the above two transits from A13 alone because our focus is the determination of ψ.

4.3 Information from Asteroseismology Analysis

In Sects. 4.4 and 4.5, we complement the analysis of the RM effect and transit light curve with the constraints on $i_\star$, $\rho_\star$, and $v \sin i_\star$ from asteroseismology to determine true obliquity ψ. This section briefly summarizes how those constraints are obtained from asteroseismic analyses; more detail is found in Sects. 3 through 5 of Benomar et al. (2014).

4.3.1 Mode Identification and Frequency Measurements

In a convective envelope of a Sun-like star, turbulent motion stochastically generates acoustic waves. While the waves gradually damp as they propagate, the oscillations with frequencies close to the eigenmodes of the star are sustained as acoustic modes. These oscillation modes can be observed in the power spectrum of the stellar light

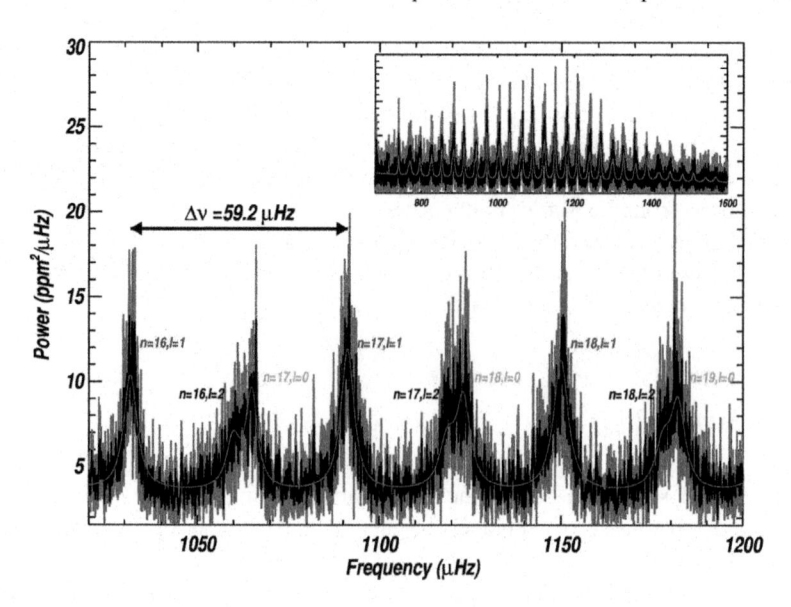

Fig. 4.2 Power spectrum of HAT-P-7 showing the three radial orders of modes with highest signal-to-noise ratio. The spectrum is shown after a boxcar smoothing over 0.08 μHz (gray) and 0.24 μHz (black). The best-fit model is the solid red line. The inset shows all the extracted modes

curves, as shown in Figs. 4.2 and 4.3 for HAT-P-7 and Kepler-25, respectively. Note that the oscillation modes cannot be seen in the time-domain (i.e., light curves), because they are not the coherent oscillations inherently to their stochastic nature of excitation.

Assuming a spherical star, each mode is labeled with three quantum numbers (n, l, m), in an analogous manner to the energy eigenstates of a hydrogen atom in quantum mechanics. The shape of each oscillation mode is given by the Lorentzian profile, whose height and width are determined by specific mechanisms of, e.g., mode excitation and damping. The frequencies of each oscillation mode can be derived by fitting this profile to the observed power spectra.

In the absence of stellar rotation, spherical symmetry assures that the frequency of each eigenmode ν depends on n and l alone. For the low angular degrees of high order modes near the surface, which satisfy $n \gg l \sim 1$, ν is almost equally spaced as

$$\nu(n, l) = \Delta\nu \left(n + \frac{l}{2} + \alpha\right) + \varepsilon_{n,l}. \tag{4.2}$$

Here $\Delta\nu$ is a characteristic frequency of the oscillation called the frequency spacing, α is a constant of order unity, and $\varepsilon_{n,l}$ is the correction related to the detailed interior structure of the star. Equation (4.2) assures that, if the power spectrum (as in Figs. 4.2 and 4.3) is divided into the chunks of width $\Delta\nu$ and lined up vertically after aligning the central frequency of each chunk, the modes (or frequency peaks) with

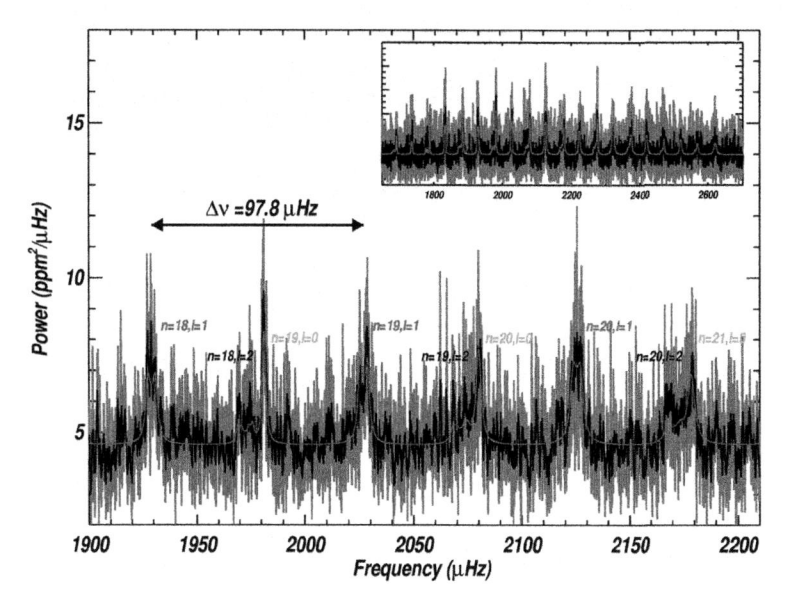

4.3 Information from Asteroseismology Analysis

Fig. 4.3 Power spectrum of Kepler-25 showing the three radial orders of modes with highest signal-to-noise ratio. The spectrum is shown after a boxcar smoothing over 0.21 μHz (gray) and 0.83 μHz (black). The best-fit model is the solid red line. The inset shows all the extracted modes

the same (n, l) should appear as nearly vertical lines, within the small correction of $\varepsilon_{n,l}$. Figures 4.4 and 4.5 created in such a way are called Échelle diagram and help the mode degree identification.

4.3.2 Derivation of Fundamental Stellar Properties

The frequency spacing $\Delta\nu$ in Eq. (4.2) is given by

$$\Delta\nu = \left(2\int_0^{R_\star} \frac{1}{c(r)} dr\right)^{-1}, \qquad (4.3)$$

where $c(r)$ is the sound speed at radius r: that is, $\Delta\nu$ is the inverse of the sound-crossing time within the star. For a star in hydrostatic equilibrium, the latter timescale is the same as the free-fall timescale, and so $\Delta\nu$ scales as the square root of the mean stellar density.[2] The scaling allows for the estimate of mean stellar density by scaling the solar values $\rho_\odot = (1.4060 \pm 0.0005) \times 10^3 \,\mathrm{kg\,m^{-3}}$ and $\Delta\nu_\odot = 135.20 \pm 0.25$ μHZ (García et al. 2011) as

---
[2] Since the pressure gradient supports the gravity, $(1/\rho)(p/R_\star) \sim c^2/R_\star \sim GM_\star/R_\star^2$ or $c/R_\star \sim \sqrt{G\rho_\star}$.

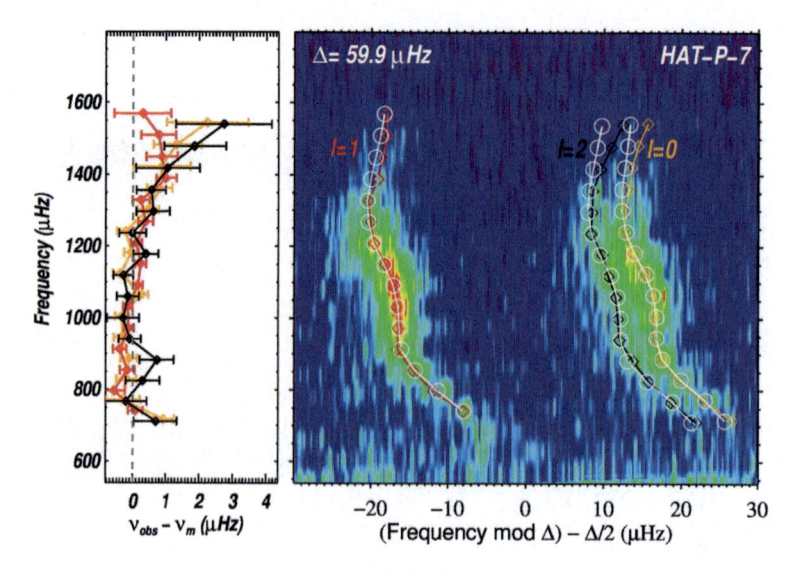

Fig. 4.4 (*Left*) Difference between the observed frequencies ν_{obs} of HAT-P-7 and the best model frequencies ν_{m}. The modes with $l = 0, 1, 2$ are shown by orange, red, and black diamonds, respectively. (*Right*) Échelle diagram showing the observed power spectrum (background), observed frequencies (diamonds), and the frequencies from the best model (white circles)

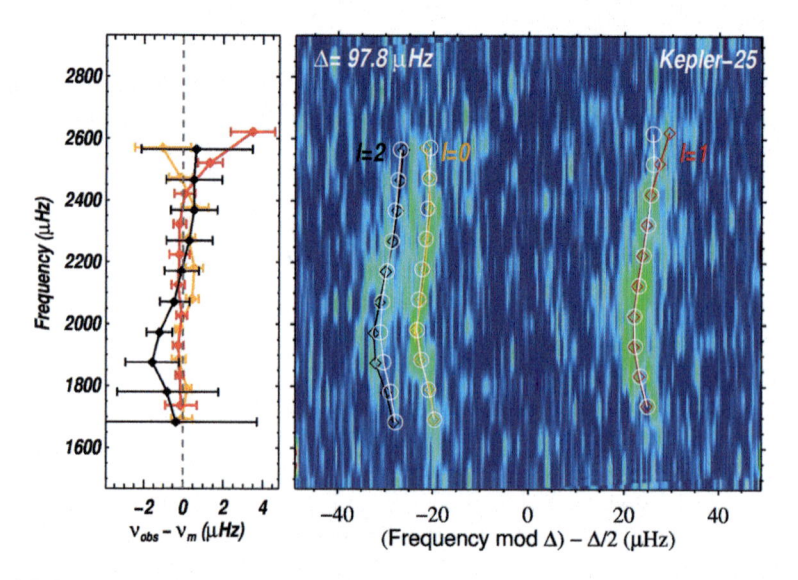

Fig. 4.5 The same as Fig. 4.4 for Kepler-25

4.3 Information from Asteroseismology Analysis

$$\rho_{\star,s} = \rho_\odot \left(\frac{\Delta\nu}{\Delta\nu_\odot}\right)^2. \quad (4.4)$$

While the scaling law (4.4) is known to hold well, this assumes that the overall internal property is similar to that of the sun, which is not the case in general. A more physically motivated (though model dependent) constraint can be obtained by fully modeling the stellar internal structure, computing the eigenfrequencies for the model, and directly comparing them to the observed oscillation frequencies. Such an analysis does not only yield model-based mean stellar density, $\rho_{\star,m}$, but also gives precise constraints on other fundamental properties of the star. They are listed in Table 4.1. In Table 4.2, we list the atmospheric parameters of the star from spectroscopy, which are also used in the above modeling of mode frequencies.

Table 4.1 Stellar parameters of HAT-P-7 and Kepler-25 derived from the modeling with the "astero" module of the Modules for Experiments in Stellar Astrophysics (MESA, Paxton et al. 2011, 2013). The mean stellar density derived from the scaling relation (4.4), $\rho_{\star,s}$, is also listed for comparison with the value from the model, $\rho_{\star,m}$

Parameter	HAT-P-7	Kepler-25
$M_\star$ ($M_\odot$)	1.59 ± 0.03	1.26 ± 0.03
$R_\star$ ($R_\odot$)	2.02 ± 0.01	1.34 ± 0.01
[Fe/H]	0.32 ± 0.04	0.11 ± 0.03
T_{eff} (K)	6310 ± 15	6354 ± 27
Age (Myr)	1770 ± 100	2750 ± 300
α_{ov}	$0.000^{+0.002}_{-0.000}$	0.007 ± 0.003
$L_\star$ ($L_\odot$)	5.84 ± 0.05	2.64 ± 0.07
$\log g$ (cgs)	4.029 ± 0.002	4.285 ± 0.003
$\rho_{\star,m}$ (10^3 kg m^{-3})	0.2708 ± 0.0035	0.7367 ± 0.0137
$\rho_{\star,s}$ (10^3 kg m^{-3})	0.2696 ± 0.0011	0.7356 ± 0.0030

Table 4.2 Non-seismic observables of HAT-P-7 and Kepler-25. All but $v \sin i_\star$ are used for stellar modeling

Parameter	HAT-P-7	Kepler-25
T_{eff} (K)	6350 ± 80	6270 ± 79
[Fe/H]	0.26 ± 0.08	-0.04 ± 0.10
$L_\star$ ($L_\odot$)	4.9 ± 1.1	...
$\log g$ (cgs)	4.070 ± 0.06	4.278 ± 0.03
$v \sin i_\star$ (km s^{-1})	3.8 ± 0.5	9.5 ± 0.5
Source	Pál et al. (2008)	Marcy et al. (2014)

4.3.3 Geometry from the Rotational Splitting

So far we have focused on the frequency information of each mode. In contrast, relative heights of the different modes tell us about the stellar rotation through the geometric effect.

In the absence of stellar rotation, the modes with the same l but with different m have the same oscillation frequencies. Stellar rotation breaks this $(2l + 1)$-fold degeneracy by splitting these modes, again analogously to the Zeeman splitting. Assuming a rigid rotation, the effect of rotational splitting is simply given by

$$\nu(n, l, m) = \nu(n, l) + m \, \delta\nu_s(n, l), \tag{4.5}$$

where the rotational splitting $\delta\nu_s(n, l)$ is the inverse of the stellar rotation period (e.g., Appourchaux et al. 2008; Benomar et al. 2009; Chaplin et al. 2013). Furthermore, the relative heights of the $2l + 1$ split modes depend on the stellar inclination through Eq. (2.3), as we described in Sect. 2.3.1. Thus, both stellar rotation and inclination can be derived by fitting the spectrum with the sum of Lorentzians with different m, weighted and shifted accordingly to Eqs. (2.3) and (4.5), respectively. The red solid lines in Figs. 4.2 and 4.3 show the best-fit spectrum models obtained in this way.

Ideally, Eqs. (2.3) and (4.5) contain enough information to specify both rotation period and stellar inclination separately. In reality, however, it is often the case for Sun-like stars as analyzed here that the splitting of the modes is not clear (see the power spectra in Figs. 4.2 and 4.3). For this reason, the amount of frequency splitting and mode amplitudes are degenerate, which produces the strong correlation between the resulting rotation frequency and inclination. This situation is clearly illustrated in Figs. 4.6 and 4.7, which show joint probability distributions of the rotation frequency and inclination of two stars.

In the joint analyses below, we use the joint probability distribution for $i_\star$ and $v \sin i_\star$ computed from the rotation period, $i_\star$, and $R_\star$ from the stellar modeling, because $v \sin i_\star$ is more directly related to the observable of the RM effect than the rotation period. We also incorporate the constraint on $\rho_\star$ as an independent Gaussian. This treatment is justified because the constraints on geometric parameters are essentially independent from those on the parameters describing the interior structure, which come from the frequency information alone.

4.3.4 Comments on the Results for Each System

In the case of HAT-P-7, splitting of the modes with different m is not clear at all, as shown in Fig. 4.2. This means that the solutions including (i) relatively fast rotation with the stellar inclination close to $0°$, and (ii) very slow rotation with an arbitrary inclination are both allowed; this explains the correlation between rotation frequency and inclination in Fig. 4.6. Since the solution (ii) has a larger volume in the param-

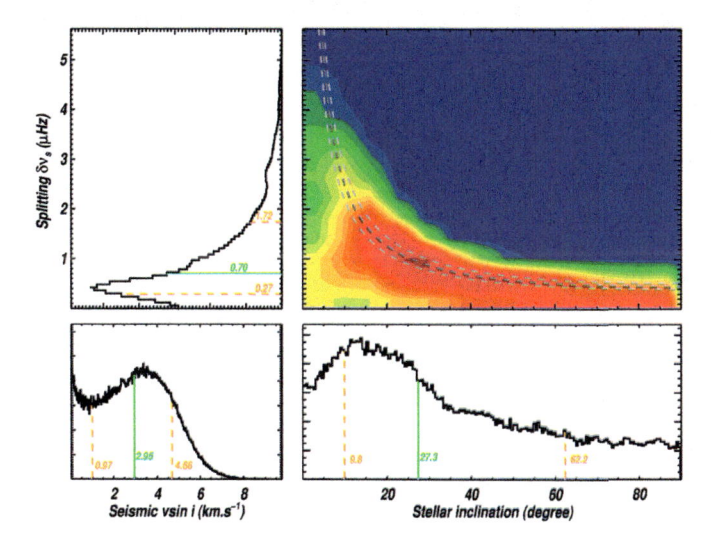

4.3 Information from Asteroseismology Analysis 65

Fig. 4.6 (*Upper right*) Joint posterior probability distribution of the stellar inclination and the rotation frequency of HAT-P-7. The red and blue colors represent the regions of the highest and lowest probabilities. The gray dotted line denotes the spectroscopic $v \sin i_\star$ from P08 with its 1σ uncertainty intervals shown with the light-gray dotted lines. (*Upper left*) Marginalized probability density function for the rotational splitting. (*Lower right*) Marginalized probability density function for the stellar inclination. (*Lower left*) Marginalized probability density function for the $v \sin i_\star$ inferred from those of the rotational splitting, stellar inclination, and stellar radius. Green and orange lines in the marginalized probability densities show the median and 68.3% credible interval, respectively

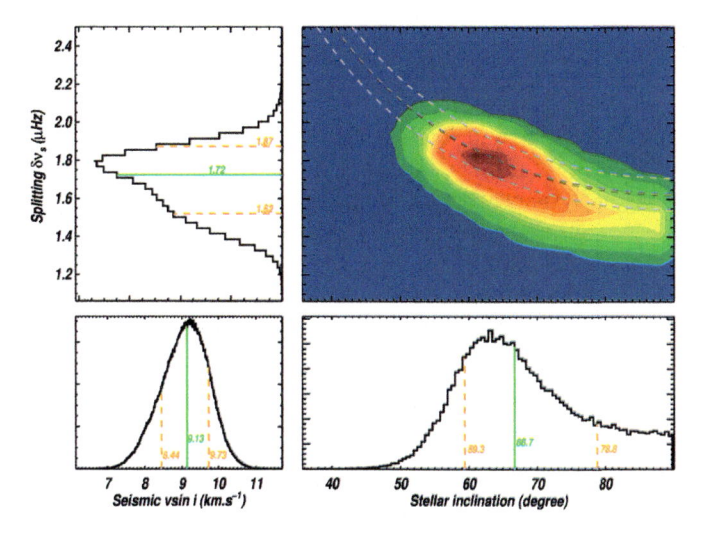

Fig. 4.7 The same as Fig. 4.6 for Kepler-25

eter space, slow rotation (i.e., small rotation frequency) is more pronounced in the marginalized posterior. In other words, it does *not* mean that the faster rotation is clearly excluded by the data.

The situation is better for Kepler-25, for which the splitting is better observed, though not clear, as perceived by the red solid lines in Fig. 4.3.

4.4 Joint Analysis of the HAT-P-7 System

In this section and the next, we combine $i_\star$ from asteroseismology and λ from the RM effect to constrain the three-dimensional spin–orbit angle ψ. Since the seismic $v \sin i_\star$ and $\rho_\star$ are also complementary to those from the RM effect and transit photometry, we reanalyze the RM effect and the whole available *Kepler* light curves simultaneously, incorporating the constraints on $i_\star$, $v \sin i_\star$, and $\rho_\star$ described in the previous sections as the prior knowledge. The method and results are presented in this section for HAT-P-7 and in the next section for Kepler-25.

For the HAT-P-7 system, the combination of asteroseismology and *Kepler* light curves provides a unique opportunity to tightly constrain the orbital eccentricity of HAT-P-7b, especially because the occultation (secondary eclipse) is clearly detected for this giant and close-in planet. Therefore, we first describe how the transit and occultation light curves constrain the planetary orbit in Sect. 4.4.1, before reporting the joint analysis for ψ in Sect. 4.4.2.

4.4.1 Analysis of Transit and Occultation Light Curves

4.4.1.1 Data Processing and Revised Ephemeris

In the following analysis, we use the *Kepler* short-cadence Pre-search Data Conditioned Simple Aperture Photometry (PDCSAP) fluxes through Q0 to Q17 retrieved from the NASA exoplanet archive.[3]

First, light curves are detrended and normalized by fitting a third-order polynomial to the out-of-transit fluxes around ±0.5 days of every transit center. Here, the central time and the duration of each transit are determined from the central time of the first observed transit calculated from the linear ephemeris, t_0, the orbital period, P, and the duration taken from the archive. We iterate the polynomial fit until all the outliers exceeding the 5σ level are excluded. In this process, we remove the transits whose baselines cannot be determined reliably due to the data gap around the ingress or egress.

Second, we fit each detrended and normalized transit with the analytic light curve model by Ohta et al. (2009) to determine its central time. We fix the planet-to-star

---

[3] http://exoplanetarchive.ipac.caltech.edu.

4.4 Joint Analysis of the HAT-P-7 System

radius ratio, $R_p/R_\star$, the ratio of the semi-major axis to the stellar radius, $a/R_\star$, the cosine of the orbital inclination, $\cos i_{\rm orb}$, at those values from the archive, adopt the coefficients for the quadratic limb-darkening law, u_1 and u_2, from Jackson et al. (2012), and assume zero orbital eccentricity (e). Since only the out-of-transit outliers were removed in the first step, we also iteratively remove in-transit outliers using the same 5σ threshold. The resulting transit times are used to phase fold all the transits and to improve the transit parameters and orbital period P.

Using these revised transit parameters, we again fit each transit light curve for its central time and total duration. Here we assume $e = 0$, fix the values of $u_1, u_2, a/R_\star$, $R_p/R_\star$, and P, and float only central transit time and $\cos i_{\rm orb}$. From these transit times, we calculate the revised ephemeris $t_0(\mathrm{BJD}) - 2454833 = 121.3585049(49)$ and $P = 2.204735427(13)$ days by linear regression. Since we find no systematic TTVs, hereafter we assume that the orbit of HAT-P-7b is described by the strictly periodic Keplerian orbit with t_0 and P obtained above.

4.4.1.2 Orbital Eccentricity and Mean Stellar Density from the Phase-Folded Transit and Occultation

The top and middle panels of Fig. 4.8 respectively show the transit and occultation light curves stacked using the revised ephemeris. The light curves are averaged into 1-min bins and the uncertainty of the flux in the i-th bin, $\sigma_{i,\mathrm{MAD}}$, is calculated as 1.4826 times median absolute deviation divided by the square root of the number of data points in the bin (Bevington 1969). Solid lines are the best-fit light curves obtained from the simultaneous fit to both light curves. We use the transit model by Mandel and Agol (2002), and binned model fluxes are calculated by averaging fluxes sampled at 0.1-min interval. In this figure, the transit and occultation are shifted in time by $t_{\rm c,tra}$ and $P/2 + t_{\rm c,tra}$, respectively, where $t_{\rm c,tra}$ is the central time of the phase-folded transit light curve. This parameter is introduced to take into account the uncertainty in t_0, and the best-fit value of $t_{\rm c,tra}$ is indeed within that uncertainty (see Table 4.3). In the transit residuals (top panel), we reproduce the anomaly first reported by Morris et al. (2013), who attributed it to the planet-induced gravity darkening. We will analyze this anomaly in Sect. 5.5.

Since the asymmetry of the planetary orbit alters the relative duration of the transit and occultation, as well as their time interval, one can tightly constrain the orbital eccentricity from the combination of transits and occultations; see Appendix B.3.3. The bottom panel of Fig. 4.8 illustrates this subtle effect by comparing the best-fit transit and occultation light curves. Here the depth of the occultation is scaled by δ, the occultation depth divided by $(R_p/R_\star)^2$, for ease of comparison. In this panel, the egress of the occultation occurs slightly later than that of the transit, while the difference is smaller for their ingresses. In other words, our best-fit model indicates that the occultation duration is longer than the transit one and that the center of occultation deviates from $P/2$. These are most likely due to the asymmetry of the orbit introduced by the slight but non-zero eccentricity, as well as the time delay of 4.5×10^{-4} days due to the finite speed of light (twice the orbital semi-major

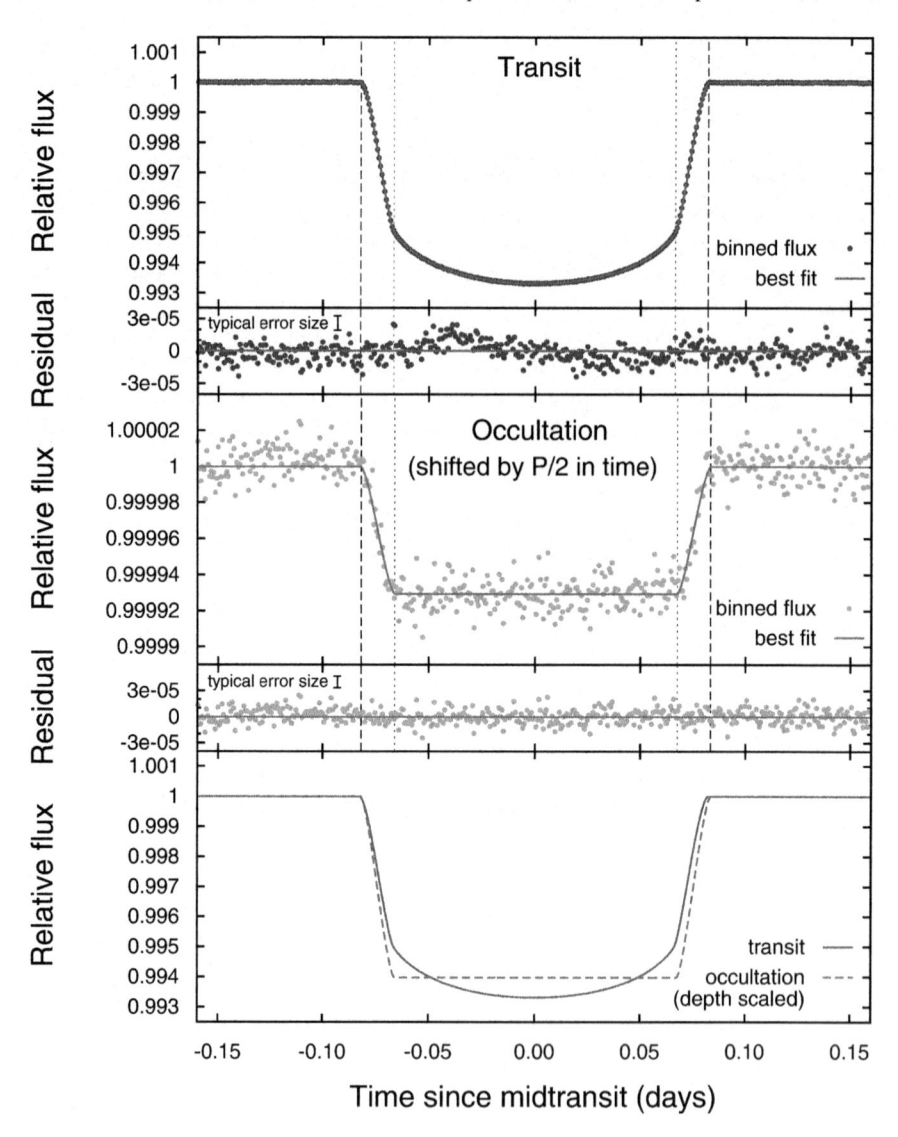

Fig. 4.8 Phase-folded transit (top) and occultation (middle) light curves. Points are the binned fluxes (1 min) and solid lines show the best-fit model light curves. Vertical dashed and dotted lines correspond to the four "contact points" where the planetary disk is tangent to the stellar limb. In the bottom panel, we compare the durations and central times of best-fit transit and occultation light curves. Occulation is shifted by $P/2$ in time in the middle and the bottom panels, and its depth is scaled by δ in the bottom panel for ease of comparison

axis divided by the speed of light; calculated for $M_\star = 1.59\,M_\odot$). In fact, with the non-zero eccentricity and the above light-travel time included, the simultaneous fit to the phase-folded transit and occultation light curves give tight constraints on the

4.4 Joint Analysis of the HAT-P-7 System 69

planet's eccentricity, $e\cos\omega = 0.00026 \pm 0.00015$ and $e\sin\omega = 0.0041 \pm 0.0022$, where ω is the argument of periastron measured from the plane of the sky.

Since $e\sin\omega$ and $a/R_\star$ are degenerate in determining the transit durations, the tight constraint on $e\sin\omega$ also allows for the accurate determination of $a/R_\star$, and hence the mean stellar density $\rho_\star$ independently from asteroseismology (Seager and Mallén-Ornelas 2003). We obtain $a/R_\star = 4.131 \pm 0.009$ from the above fit, and then derive $\rho_\star = (0.275 \pm 0.002) \times 10^3$ kg m^{-3} from Kepler's third law,

$$\rho_\star = \frac{3\pi}{GP^2} \left(\frac{a}{R_\star}\right)^3 \left(1 + \frac{M_p}{M_\star}\right)^{-1}, \qquad (4.6)$$

where G denotes the gravitational constant, and $M_p/M_\star \sim 10^{-3}$ can be neglected. This value is larger than $\rho_{\star,s}$ based on the seismic scaling relation by $2.4\,\sigma$, but consistent with $\rho_{\star,m}$ from the stellar model at the $1\,\sigma$ level (see Table 4.1). For this reason, we adopt the constraints from the stellar model as the prior information in the following joint fit. The choice of the prior, however, does not affect the spin–orbit angle determination, but only slightly changes the values of $a/R_\star$, $\rho_\star$, $\cos i_{\rm orb}$, and $e\sin\omega$. The slight discrepancy between $\rho_\star$ from the seismic scaling relation ($\rho_{\star,s}$) and that from transit and occultation implies that the current precision of the *Kepler* photometry even enables an independent test of the seismic scaling relation for the mean stellar density.

4.4.2 Joint Analysis

4.4.2.1 Method

In this subsection, we report the joint MCMC analysis of phase-folded transit and occultation light curves (cf. Sect. 4.4.1) and RVs (cf. Sect. 4.2.1) making use of the prior constraints on the mean stellar density $\rho_\star$, projected stellar rotational velocity $v \sin i_\star$, and stellar inclination $i_\star$ obtained from asteroseismology in Sect. 4.3. As discussed in Sect. 4.4.1, the precise constraint on $\rho_\star$ (equivalent to that on $a/R_\star$) helps to lift the degeneracy between $a/R_\star$ and $e\sin\omega$, thus resulting in improved constraints on these two parameters. In addition, $v \sin i_\star$ is the key parameter for the RM effect along with λ, and so the constraint on $v \sin i_\star$ helps us to better determine λ from the observed RM signal. Finally, $i_\star$ is crucial in determining the three-dimensional spin–orbit angle ψ via Eq. (4.1), which is the major goal of this chapter.

In order to properly handle the possible correlation among λ, $v \sin i_\star$, and $i_\star$, we adopt the joint probability distribution for $v \sin i_\star$ and $i_\star$ as the prior in our MCMC analysis and directly calculate the posterior distribution for ψ by floating $i_\star$ as well. It should be noted here that our observables do not determine the sign of $\cos i_\star$ or $\cos i_{\rm orb}$, due to the symmetry with respect to the plane of the sky. In order to take into account this inherent degeneracy, we randomly change the sign of the first term in Eq. (4.1) in computing ψ. Since the probability distribution of $\rho_\star$ is almost independent of those of $v \sin i_\star$ and $i_\star$, we include the constraint on this parameter as an independent Gaussian with the central value and width of $\rho_{\star,m}$ listed in Table 4.1.

We adopt the same model (including non-zero eccentricity and light-travel time) for transit and occultation as in Sect. 4.4.1. The observed RVs are modeled as

$$v_{\star,\text{model}}(t) = v_{\star,\text{orb}}(t) + v_{\star,\text{RM}}(t) + \gamma_i + \dot{\gamma}(t - t_0). \qquad (4.7)$$

Here,

$$v_{\star,\text{orb}} = K_\star [\cos(\omega + f) + e \cos \omega] \qquad (4.8)$$

is the stellar orbital RVs for the Keplerian orbit, where $K_\star$ is the RV semi-amplitude (cf. Eq. 1.4) and f is the true anomaly of the planet. The γ_i ($i = 1, 2$) are the constant offsets for RVs from Keck/HIRES ($i = 1$) and Subaru/HDS ($i = 2$), and $\dot{\gamma}$ accounts for the linear trend in the observed RVs in the W09 data set (Winn et al. 2009; Narita et al. 2012; Knutson et al. 2014). Finally, anomalous RVs due to the RM effect, $v_{\star,\text{RM}}$, are modeled using the analytic formula by Hirano et al. (2011). The parameters characterizing the RM model include $v \sin i_\star$ (projected rotational velocity of the star), β (Gaussian dispersion of spectral lines), γ (Lorentzian dispersion of spectral lines), ζ (macroturbulence dispersion of spectral lines), $u_{1\text{RM}} + u_{2\text{RM}}$, and $u_{1\text{RM}} - u_{2\text{RM}}$ (coefficients for the quadratic limb-darkening law in the RM effect). We do not take into account the effect of convective blueshift (Shporer and Brown 2011), as its typical amplitude (~ 1 m s^{-1}) is smaller than the precision of the RVs analyzed here.

We impose the non-seismic priors as well on some of the model parameters. For the ephemeris, we use the Gaussian priors t_0(BJD) $- 2454833 = 121.3585049 \pm 0.0000049$ and $P = 2.204735427 \pm 0.000000013$ days obtained from the transit light curves. The priors on the RM parameters (β, γ, ζ, $u_{1\text{RM}} + u_{2\text{RM}}$, and $u_{1\text{RM}} - u_{2\text{RM}}$) are almost the same as in A12. Namely, we fix $\beta = 3$ km s^{-1} and $\gamma = 1$ km s^{-1}, and assume Gaussian prior $\zeta = 5.18 \pm 1.5$ km s^{-1}. We fix the value of $u_{1\text{RM}} - u_{2\text{RM}}$ at -0.023 from the tables of Claret (2000) for the Johnson V band and the ATLAS model. The value is obtained using the jktld tool[4] for the parameters $T_{\text{eff}} = 6350$ K, $\log g$ (cgs) $= 4.07$, and [Fe/H] $= 0.3$. The value of $u_{1\text{RM}} + u_{2\text{RM}}$ is floated around the tabulated value of 0.70 assuming the Gaussian prior of width 0.10. In addition, we impose an additional Gaussian prior on $v \sin i_\star$ based on the spectroscopic value in Table 4.2, because the seismic constraint on this parameter is independent of the spectroscopic $v \sin i_\star$. We assume uniform priors for the other 13 fitting parameters listed in Table 4.3 (top and middle blocks).

In the joint fit, we assume the same values of stellar jitter as used in the original papers; 9.3 m s^{-1} for the W09 set, 3.8 m s^{-1} for the Keck/HIRES RVs of the N09 set, and 6.0 m s^{-1} for the A12 set. In order to prevent the transit and occultation light curves from placing unreasonably tight constraints compared to RVs, we

---

[4] http://www.astro.keele.ac.uk/jkt/codes/jktld.html.

4.4 Joint Analysis of the HAT-P-7 System

also increase the errors quoted for photometric data (evaluated in Sect. 4.4.1.2) as $\sigma_i = \sqrt{\sigma_{i,\text{MAD}}^2 + \sigma_r^2}$. Here, $\sigma_r = 5.8 \times 10^{-6}$ is a parameter analogous to the RV jitter and chosen so that the reduced χ^2 of the light curve fit becomes unity. This prescription is also motivated by the following two facts. First, $\sigma_{i,\text{MAD}}$ tends to underestimate the true uncertainty because it neglects the effect of correlated noise. Indeed, when the number of data points is sufficiently large, uncertainties are dominated by the correlated or "red" noise component (Pont et al. 2006). Second, the systematic residuals of the best-fit transit model (top panel of Fig. 4.8) suggest other effects that are not taken into account in our model (see Sect. 5.5 for the detailed analysis of this feature). Placing too much weights on such features could bias the transit parameters.

4.4.2.2 Results

Constraints on the system parameters from the joint analysis are summarized in Table 4.3. The corresponding joint posterior distributions are shown in Figs. C.1 through C.3 in Appendix C to elucidate the parameter correlations. The "parameters mainly derived from light curves/RVs" are the model (fitted) parameters, while the "derived quantities" are the parameters derived from the fitted parameters (along with $M_\star$ and $R_\star$ in Table 4.1 for M_p, R_p, and ρ_p). While our result is in a reasonable agreement with previous studies (cf. Morris et al. 2013; Esteves et al. 2013; Van Eylen et al. 2013), it provides two major improvements.

First, we determine the orbital eccentricity of HAT-P-7b essentially from the photometry (i.e., transit, occultation, and asteroseismology) alone. A similar method has recently been employed by Van Eylen et al. (2014) to constrain the planet's orbital eccentricity using the seismic stellar density (see also Dawson and Johnson 2012; Kipping 2014), but here we show that this method is also useful for such a low-eccentricity orbit. Furthermore, our result is even more precise and reliable because it takes into account the independent constraint on $\rho_\star$ and e from the occultation light curve.

Second, we obtain the probability distribution for the true obliquity ψ, rather than the sky-projected one λ, in a consistent manner. In the case of HAT-P-7, the constraint on ψ is not very strong because the modest splitting of the azimuthal modes only allows a weak constraint on $i_\star$ (see Fig. 4.6). Nevertheless, we find that the peak values of ψ shift towards 90° compared to those obtained from the "random" $i_\star$ uniform in $\cos i_\star$ (i.e., without the knowledge from asteroseismology) in all three data sets, as shown in Fig. 4.9. Moreover, the methodology presented here can be applied to other systems, for some of which asteroseismology may be able to tightly constrain $i_\star$ unlike HAT-P-7. We will show that this is indeed the case for the Kepler-25 system in the next section.

Table 4.3 Parameters of the HAT-P-7 System from the Joint Analysis

Parameter	Value (W09)	Value (N09)	Value (A12)
Parameters mainly derived from light curves (transit, occultation, asteroseismology)			
t_0 (BJD) $-$ 2454833		$121.3585049 \pm 0.0000049$	
P (day)		$2.204735427 \pm 0.000000013$	
$e \cos \omega$	0.00024 ± 0.00020	0.00024 ± 0.00020	0.00025 ± 0.00020
$e \sin \omega$	$0.0053^{+0.0022}_{-0.0021}$	$0.0057^{+0.0025}_{-0.0026}$	$0.0049^{+0.0026}_{-0.0030}$
u_1	0.3540 ± 0.0034	$0.3544^{+0.0033}_{-0.0034}$	$0.3545^{+0.0034}_{-0.0035}$
u_2	$0.1670^{+0.0055}_{-0.0054}$	$0.1663^{+0.0055}_{-0.0053}$	$0.1661^{+0.0056}_{-0.0055}$
$\rho_\star$ (10^3 kg m^{-3})	0.2736 ± 0.0016	$0.2731^{+0.0021}_{-0.0018}$	$0.2737^{+0.0024}_{-0.0018}$
$\cos i_{\rm orb}$	$0.12149^{+0.00056}_{-0.00057}$	$0.12166^{+0.00063}_{-0.00068}$	$0.12145^{+0.00061}_{-0.00081}$
$R_{\rm p}/R_\star$	$0.077589^{+0.000020}_{-0.000021}$	0.077593 ± 0.000020	$0.077591^{+0.000020}_{-0.000021}$
δ		0.01171 ± 0.00010	
$t_{\rm c,tra}$ (day)		$-0.0000044^{+0.0000041}_{-0.0000042}$	
$i_\star$ (°)	31^{+33}_{-16}	33^{+34}_{-20}	33^{+34}_{-20}
Parameters mainly derived from RVs			
$K_\star$ (m s^{-1})	211.7 ± 2.3	213.2 ± 1.8	214.0 ± 4.6
γ_1 (m s^{-1})	-15.5 ± 3.0	-37.5 ± 1.5	$10.4^{+1.5}_{-1.6}$
γ_2 (m s^{-1})	-9.7 ± 1.7	-16.9 ± 1.4	–
$\dot{\gamma}$ (m s^{-1} yr^{-1})	21.5 ± 2.5	–	–
λ (°)	186^{+10}_{-11}	$220.3^{+8.2}_{-9.3}$	157^{+14}_{-13}
$v \sin i_\star$ (km s^{-1})	$4.15^{+0.38}_{-0.39}$	3.17 ± 0.33	$3.17^{+0.33}_{-0.34}$
β (km s^{-1})		3.0 (fixed)	
γ (km s^{-1})		1.0 (fixed)	
ζ (km s^{-1})	5.3 ± 1.5	5.5 ± 1.5	5.5 ± 1.5
$u_{1\rm RM} + u_{2\rm RM}$		0.70 ± 0.10	
$u_{1\rm RM} - u_{2\rm RM}$		-0.23 (fixed)	
Derived quantities			
ψ (°)	122^{+30}_{-18}	115^{+19}_{-16}	120^{+26}_{-18}
$a/R_\star$	$4.1269^{+0.0082}_{-0.0078}$	$4.1245^{+0.0103}_{-0.0092}$	$4.1277^{+0.0121}_{-0.0090}$
Impact parameter of transit ($R_\star$)	0.4987 ± 0.0013	0.4989 ± 0.0013	$0.4988^{+0.0013}_{-0.0014}$
$T_{14,\rm tra}$ (day)	0.164301 ± 0.000022	0.164303 ± 0.000023	0.164300 ± 0.000023
$T_{23,\rm tra}$ (day)	$0.133042^{+0.000049}_{-0.000048}$	$0.133034^{+0.000047}_{-0.000048}$	$0.133037^{+0.000052}_{-0.000048}$
$T_{\rm tra}$ (day)	$0.148672^{+0.000025}_{-0.000024}$	0.148668 ± 0.000024	$0.148669^{+0.000025}_{-0.000024}$
Impact parameter of occultation ($R_\star$)	$0.5040^{+0.0022}_{-0.0023}$	$0.5047^{+0.0025}_{-0.0028}$	$0.5039^{+0.0024}_{-0.0033}$
$T_{14,\rm occ}$ (day)	$0.16555^{+0.00051}_{-0.00050}$	$0.16566^{+0.00058}_{-0.00061}$	$0.16547^{+0.00060}_{-0.00070}$
$T_{23,\rm occ}$ (day)	$0.13385^{+0.00034}_{-0.00033}$	$0.13392^{+0.00039}_{-0.00040}$	$0.13379^{+0.00041}_{-0.00046}$
$T_{\rm occ}$ (day)	$0.14970^{+0.00042}_{-0.00041}$	$0.14979^{+0.00048}_{-0.00051}$	$0.14963^{+0.00050}_{-0.00058}$

(continued)

Table 4.3 (continued)

Parameter	Value (W09)	Value (N09)	Value (A12)
Derived quantities			
Occultation depth (ppm)		70.5 ± 0.6	
$M_p (M_J)$	1.86 ± 0.03	1.87 ± 0.03	1.88 ± 0.05
$R_p (R_J)$		1.526 ± 0.008	
ρ_p (10^3 kg m^{-3})	0.65 ± 0.01	0.66 ± 0.01	0.66 ± 0.02

Note The quoted best-fit values are the medians of their MCMC posteriors, and uncertainties exclude 15.87% of values at upper and lower extremes. The T_{ij} ($i, j = 1, 2, 3, 4$) is the duration between the two contact points i and j [see Fig. 2 of Winn (2011) for their definitions], and $T = (T_{14} + T_{23})/2$. The subscript "tra" refers to transits and "occ" to occultations

4.5 Joint Analysis of the Kepler-25 System

4.5.1 Method

We repeat almost the same analysis for Kepler-25c as in Sect. 4.4. There are, however, several differences in the light curve and RV analyses as described below, mainly due to the multiplicity of the Kepler-25 system and relatively small signal-to-noise ratio of the Kepler-25c's transit:

1. We phase-fold the transits using the actually observed transit times rather than those calculated from the linear ephemeris. This is because the transit times of Kepler-25c ($P = 12.7$ days) exhibit significant TTVs due to the proximity to the 2 : 1 mean-motion resonance with Kepler-25b ($P = 6.2$ days). This is why we do not allow $t_{c, \text{tra}}$, the central time of the phase-folded transit, to be a free parameter. We adopt $\sigma_r = 1.6 \times 10^{-5}$ based on the χ^2 of the light curve fit.
2. The occultation of Kepler-25c was not detected and not taken into account in the following analysis.
3. As the quality of the transit light curve of Kepler-25c is not so good as that of HAT-P-7b, we could not determine the limb-darkening coefficients very well. For this reason, we impose the prior $u_1 - u_2 = -0.0015 \pm 0.50$ based on the tables of Claret (2000), and choose $u_1 + u_2$ and $u_1 - u_2$, instead of u_1 and u_2, as free parameters. We made sure that the choice of the prior width for $u_1 - u_2$ does not affect the constraint on ψ.
4. In order to take into account the other planets in the RV fit, we allow the orbital semi-amplitude $K_\star$ and RV offset γ for each of the nights in 2011 and 2012 to be free parameters, as in A13. RV jitters are fixed at 3.3 m s^{-1}.
5. We do not fit the orbital eccentricity but fix $e = 0$, because we do not analyze the occultation nor RVs throughout the orbit (Marcy et al. 2014).
6. We assume the independent Gaussian priors $u_{1\text{RM}} + u_{2\text{RM}} = 0.69 \pm 0.10$ and $\zeta = 4.85 \pm 1.5$ km s^{-1} from A13, and fix $u_{1\text{RM}} - u_{2\text{RM}} = -0.0297$ from the tables of Claret (2000).

Fig. 4.9 Probability distributions for the three-dimensional spin–orbit angle ψ of HAT-P-7b for the W09 (top), N09 (middle), and A12 (bottom) data sets. Solid red lines show the posteriors from the joint analysis, while the black ones are the probability distributions obtained from uniform $\cos i_\star$ and the posteriors of λ and i_{orb} from the joint analysis (Table 4.3). The median, 1σ lower limit, and 1σ upper limit for each distribution are shown with vertical dotted lines. A small bump around $\psi \approx 95°$ in each panel originates from the fact that each posterior shown here is the superposition of the two inherently degenerate configurations with the opposite signs of $\cos i_\star \cos i_{\mathrm{orb}}$; see the discussion in the second paragraph of Sect. 4.4.2.1

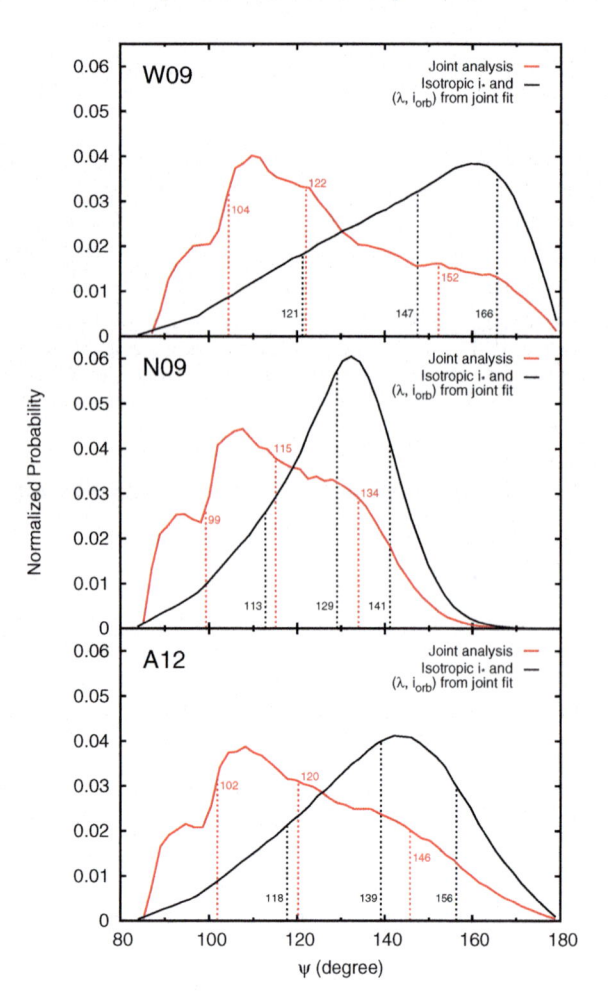

4.5.2 Results

In the case of the Kepler-25 system, the uncertainty in ψ is significantly reduced by virtue of the seismic information. This situation is clearly illustrated in Fig. 4.10, which compares the posterior probability distribution for ψ from the joint fit (solid red line) to that based on λ and i_{orb} from the joint fit and the uniform $\cos i_\star$ (solid black line). The corresponding system parameters are summarized in Table 4.4, and the joint posterior distribution can be found in Fig. C.4. They are basically consistent with those obtained by A13, except for the increased precision in the transit parameters.

Interestingly, our result suggests a spin–orbit misalignment for Kepler-25c with more than 2σ significance. In order to check the robustness of this result, we also calculate the probability distribution of ψ for the seismic $i_\star$ and an independent

4.5 Joint Analysis of the Kepler-25 System

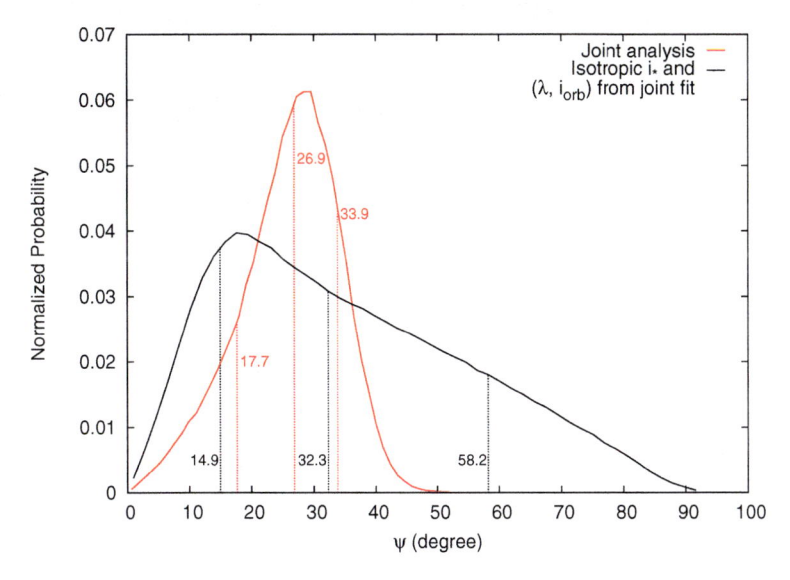

Fig. 4.10 Probability distributions for the three-dimensional spin–orbit angle ψ of Kepler-25c. The solid red line shows the posterior from the joint analysis, while the black one is the probability distribution obtained from λ and $i_{\rm orb}$ in Table 4.4 and uniform $\cos i_\star$. The median, 1 σ lower limit, and 1 σ upper limit for each distribution are shown with vertical dotted lines

Gaussian $\lambda = -0°\!.5 \pm 5°\!.7$ from the Doppler tomography. We obtain $\psi = 23°\!.7^{+8°\!.0}_{-11°\!.3}$ in this case, which still points to the spin–orbit misalignment marginally. If confirmed, this will be the first example of the spin–orbit misalignment in the multi-transiting system around a main-sequence star.[5] The implication of this result will be discussed in Sect. 4.6.2, along with some caveats in Sect. 4.6.3.

4.6 Summary and Discussion

4.6.1 HAT-P-7

From asteroseismology alone, we obtain $i_\star = 27°{}^{+35°}_{-18°}$ for HAT-P-7 (Fig. 4.6). This constraint, combined with the *Kepler* light curves and the three independent RM measurements, yields $\psi = 122°{}^{+30°}_{-18°}$ and $i_\star = 31°{}^{+33°}_{-16°}$, $\psi = 115°{}^{+19°}_{-16°}$ and $i_\star = 33°{}^{+34°}_{-20°}$, and $\psi = 120°{}^{+26°}_{-18°}$ and $i_\star = 33°{}^{+34°}_{-20°}$ for the RVs from W09, N09, and A12, respectively (Fig. 4.9 and Table 4.3). Although the resulting constraints are not very strong due to the modest splittings of azimuthal modes (see Fig. 4.6), our results

---

[5]The first spin–orbit misalignment in the multi-transiting system was confirmed by Huber et al. (2013) around a red giant star Kepler-56 using asteroseismology, as mentioned in Sects. 2.4.3 and 3.4.

Table 4.4 Parameters of the Kepler-25 System from the Joint Analysis

Parameter	Value (A13)
Parameters mainly derived from light curves (transit, asteroseismology)	
t_0(BJD) − 2454833	$127.646558^{+0.000096}_{-0.000094}$
P (day)	$12.7203724^{+0.0000014}_{-0.0000013}$
$u_1 + u_2$	0.550 ± 0.018
$u_1 - u_2$	-0.27 ± 0.44
$\rho_\star$ (10^3 kg m^{-3})	$0.733^{+0.013}_{-0.012}$
$\cos i_{\rm orb}$	$0.04788^{+0.00036}_{-0.00038}$
$R_{\rm p}/R_\star$	$0.03590^{+0.00054}_{-0.00046}$
$i_\star$ (°)	$65.4^{+10.6}_{-6.4}$
Parameters mainly derived from RVs	
$K_{\star,2011}$ (m s^{-1})	-13 ± 22
$K_{\star,2012}$ (m s^{-1})	-37 ± 30
γ_{2011} (m s^{-1})	-3.5 ± 1.3
γ_{2012} (m s^{-1})	2.0 ± 1.4
λ (°)	9.4 ± 7.1
$v \sin i_\star$ (km s^{-1})	$9.34^{+0.37}_{-0.39}$
β (km s^{-1})	3.0 (fixed)
γ (km s^{-1})	1.0 (fixed)
ζ (km s^{-1})	4.9 ± 1.5
$u_{1\rm RM} + u_{2\rm RM}$	0.69 ± 0.10
$u_{1\rm RM} - u_{2\rm RM}$	-0.0297 (fixed)
Derived quantities	
ψ (°)	$26.9^{+7.0}_{-9.2}$
$a/R_\star$	18.44 ± 0.11
Transit impact parameter ($R_\star$)	0.8826 ± 0.0018
$T_{14,\rm tra}$ (day)	0.11925 ± 0.00025
$T_{23,\rm tra}$ (day)	$0.08528^{+0.00065}_{-0.00069}$
$T_{\rm tra}$ (day)	$0.10226^{+0.00036}_{-0.00037}$

Note The quoted best-fit values are the medians of their MCMC posteriors, and uncertainties exclude 15.87% of values at upper and lower extremes. The T_{ij} ($i, j = 1, 2, 3, 4$) is the duration between the two contact points i and j [see Fig. 2 of Winn (2011) for their definitions], and $T = (T_{14} + T_{23})/2$. The subscript "tra" refers to transits and "occ" to occultations.

suggest that the orbit of HAT-P-7b is closer to the polar configuration rather than retrograde as λ may imply.

It is worth noting that the suggested discrepancies in λ and $v \sin i_\star$ in three data sets (cf. Sect. 4.2.1) still persist in our analysis. For a fair comparison with the A12 result, we repeat the same analyses for the W09 and N09 data only including RVs taken over the same night, but the values of λ and $v \sin i_\star$ do not change significantly. Since we have used the same model of the RM effect and the same priors from the

Kepler photometry for the three sets of data, our results confirm that the discrepancy comes from the RV data themselves. As A12 discussed, such a discrepancy may originate from some physics that is not included in the current model of the RM effect, but its origin is beyond the scope of this chapter.

As a by-product of the spin–orbit analysis, we have found that HAT-P-7b has a small but non-zero orbital eccentricity, $e = 0.005 \pm 0.001$ (weighted mean of the three data sets), which is consistent with $e = 0.0055^{+0.007}_{-0.0033}$ obtained by Knutson et al. (2014). Our constraint on e comes from the duration and mid-time of the occultation of HAT-P-7b relative to those of the transit, along with the constraint on the mean stellar density $\rho_\star$ from asteroseismology. This approach is justified by the fact that $\rho_\star$ from the transit and occultation alone shows a reasonable agreement with the model stellar density $\rho_{\star,\mathrm{m}}$ derived independently from asteroseismology. The origin of this non-zero e may deserve further theoretical consideration because the tides are expected to damp e rapidly for such a close-in planet as HAT-P-7b.

4.6.2 Kepler-25

For Kepler-25, we obtain $i_\star = 65°.4^{+10°.6}_{-6°.4}$ from the joint analysis. The constraint is slightly better than $i_\star = 66°.7^{+12°.1}_{-7°.4}$ from asteroseismology alone (Fig. 4.7), mainly due to the prior on $v \sin i_\star$ from spectroscopy. The constraint on $i_\star$ is better than HAT-P-7 despite the lower signal-to-noise ratio of the oscillation spectrum, because of the greater rotational splitting (see Fig. 4.7). This allows us to tightly constrain the spin–orbit angle of Kepler-25c as $\psi = 26°.9^{+7°.0}_{-9°.2}$ (Fig. 4.10). Our finding is important in two aspects: (1) this is the first quantitative measurement of ψ, instead of λ, for multi-planetary systems, except for the solar system. (2) Kepler-25 is the first system that exhibits a possible spin–orbit misalignment among the multi-transiting systems with a main-sequence host star, while it is the second example if we consider the system with a red-giant host star, Kepler-56.

The spin–orbit misalignment, if real, is particularly interesting because it may be evidence for the initial star–disk misalignment, as discussed in Chaps. 2 and 3. In this context, the orbital inclinations of the other two planets (Kepler-25b and Kepler-25d) relative to that of Kepler-25c would be of interest to further test whether the misalignment is primordial or not. They may be constrained from the analysis of TTVs and transit duration variations, combined with orbital RVs to constrain the orbit of the outer non-transiting planet d. In this chapter, we did not model these phenomena because our main concern is the determination of the spin–orbit misalignment.

It is also interesting to note that both HAT-P-7 and Kepler-25 are relatively hot stars with $T_{\mathrm{eff}} \gtrsim 6300\,\mathrm{K}$ and in line with the observed trend that the spin–orbit misalignments are preferentially found around hot stars (Sect. 2.4). Although Rogers et al. (2012) suggested that temporal variations of the stellar rotation due to internal gravity waves could explain this empirical trend (Sect. 3.3.1), we found no evidence to support this scenario for the two systems. Regarding HAT-P-7, we compared the rotational splitting from Fig. 4.6 with that from Q0 to Q2 (results from the study of Oshagh et al. 2013), but found no evidence of significant variations. Although results using only Q0 to Q2 have large uncertainties, this may indicate that the rotation remains constant over time. Moreover, we tightly constrained the rotation of Kepler-25 and showed that outer layers certainly rotate at constant velocity. This is incompatible with the scenario suggested by Rogers et al., (2012), which predicts the radial differential rotation.

4.6.3 Note on the Result for Kepler-25

After the results in this chapter were published in Benomar et al. (2014), Campante et al. (2016) independently performed a similar analysis for the sample of *Kepler* stars including HAT-P-7 and Kepler-25, but adopting a different procedure for generating the light curve from the original photometry data. While their results are consistent with ours, they found $i_\star$ peaked closer to 90° for Kepler-25, thus obtaining $\psi = 12°.6^{+6°.7}_{-11°.0}$ rather consistent with a spin–orbit alignment. Campante et al. (2016) also found an opposite shift for the multi-planet host Kepler-50; a slight misalignment like we found for Kepler-25 is favored in their analysis, while the previous study by Chaplin et al. (2013) found $i_\star$ peaked around 90°. These examples show that the current asteroseismic inference of $i_\star$ is susceptible to systematics associated with the data processing, and the results of marginal significance, including ours for Kepler-25, need to be taken with care.

4.7 Conclusion

The major purpose of the present chapter is two-fold. The first is to develop and describe a detailed methodology of determining the three-dimensional spin–orbit angle ψ for transiting planetary systems. The other is to demonstrate the power of the methodology by applying it to the two specific systems, HAT-P-7 and Kepler-25.

We find a near-polar orbit for HAT-P-7b, rather than a counter-orbiting one as naively expected from the observed $\lambda \approx 180°$. The result implies that the orbit of HAT-P-7b could naturally be formed within the current framework of high-eccentricity migration, as discussed in Sect. 3.2.1. It will be of interest to apply similar analyses to systems with measured $\lambda \approx 180°$ to test whether any of them indeed has $\psi \approx 180°$.

The true obliquity ψ of Kepler-25 is constrained for the first time in a multi-transiting system. The determination of ψ is important for multi-transiting planetary systems, where all the planets are supposed to share the same orbital plane; a large ψ in such a system indicates that the stellar spin was significantly tilted with respect to the protoplanetary disk plane, which would eventually become the orbital planes of the planets. While we find tentative evidence for such a primordial misalignment (cf. Sect. 3.3), the asteroseismic inference is currently not robust for this system and further investigation is required for a more decisive conclusion.

In addition to the determination of ψ, the joint analysis improves the accuracy and precision of numerous system parameters of a specific target. In turn, any discrepancy among the separate analyses points to a certain physical process that needs to be taken into account in the detailed modeling. Such analyses would therefore open a new window for the exploration of the origin and evolution of planetary systems.

References

C. Aerts, J. Christensen-Dalsgaard, D.W. Kurtz, *Asteroseismology* (2010)
S. Albrecht, J.N. Winn, J.A. Johnson et al., ApJ **757**, 18 (A12) (2012)
S. Albrecht, J.N. Winn, G.W. Marcy et al., ApJ **771**, 11 (A13) (2013)
T. Appourchaux, E. Michel, M. Auvergne et al., A&A **488**, 705 (2008)
O. Benomar, T. Appourchaux, F. Baudin, A&A **506**, 15 (2009)
O. Benomar, K. Masuda, H. Shibahashi, Y. Suto, PASJ **66**, 94 (2014)
P.R. Bevington, *Data Reduction And Error Analysis For The Physical Sciences* (1969)
T.L. Campante, M.N. Lund, J.S. Kuszlewicz et al., ApJ **819**, 85 (2016)
W.J. Chaplin, R. Sanchis-Ojeda, T.L. Campante et al., ApJ **766**, 101 (2013)
A. Claret, A&A **363**, 1081 (2000)
R.I. Dawson, J.A. Johnson, ApJ **756**, 122 (2012)
L.J. Esteves, E.J.W. De Mooij, R. Jayawardhana, ApJ **772**, 51 (2013)
R.A. García, D. Salabert, J. Ballot et al., J. Phys. Conf. Ser. **271**, 012049 (2011)
T. Hirano, Y. Suto, A. Taruya et al., ApJ **709**, 458 (2010)
T. Hirano, Y. Suto, J.N. Winn et al., ApJ **742**, 69 (2011)
D. Huber, J.A. Carter, M. Barbieri et al., Science **342**, 331 (2013)
B.K. Jackson, N.K. Lewis, J.W. Barnes et al., ApJ **751**, 112 (2012)
D.M. Kipping, MNRS **440**, 2164 (2014)
H.A. Knutson, B.J. Fulton, B.T. Montet et al., ApJ **785**, 126 (2014)
K. Mandel, E. Agol, ApJ **580**, L171 (2002)
G.W. Marcy, H. Isaacson, A.W. Howard et al., ApJs **210**, 20 (2014)
D.B. McLaughlin, ApJ **60**, 22 (1924)
B.M. Morris, A.M. Mandell, D. Deming, ApJ **764**, L22 (2013)
N. Narita, B. Sato, T. Hirano, M. Tamura, PASJ **61**, L35 (N09) 2009
N. Narita, Y.H. Takahashi, M. Kuzuhara et al., PASJ **64**, L7 (2012)
Y. Ohta, A. Taruya, Y. Suto, ApJ **622**, 1118 (2005)
Y. Ohta, A. Taruya, Y. Suto, ApJ **690**, 1 (2009)
M. Oshagh, A. Grigahcène, O. Benomar et al., in *Astrophysics and Space Science Proceedings, vol. 31, Stellar Pulsations: Impact of New Instrumentation and New Insights*, ed. by J.C. Suárez, R. Garrido, L.A. Balona, J. Christensen-Dalsgaard (2013), p. 227
A. Pál, G.Á. Bakos, G. Torres et al., ApJ **680**, 1450 (P08) 2008
B. Paxton, L. Bildsten, A. Dotter et al., ApJS **192**, 3 (2011)
B. Paxton, M. Cantiello, P. Arras et al., ApJS **208**, 4 (2013)
F. Pont, S. Zucker, D. Queloz, MNRS **373**, 231 (2006)
D. Queloz, A. Eggenberger, M. Mayor et al., A&A **359**, L13 (2000)
T.M. Rogers, D.N.C. Lin, H.H.B. Lau, ApJ **758**, L6 (2012)
R.A. Rossiter, ApJ **60**, 15 (1924)
S. Seager, G. Mallén-Ornelas, ApJ **585**, 1038 (2003)
A. Shporer, T. Brown, ApJ **733**, 30 (2011)

J.H. Steffen, D.C. Fabrycky, E.B. Ford et al., MNRS **421**, 2342 (2012)
W. Unno, Y. Osaki, H. Ando, H. Saio, H. Shibahashi, *Nonradial Oscillations Of Stars* (1989)
V. Van Eylen, M. Lindholm Nielsen, B. Hinrup, B. Tingley, H. Kjeldsen, ApJ **774**, L19 (2013)
V. Van Eylen, M.N. Lund, V. Silva Aguirre et al., ApJ **782**, 14 (2014)
J.N. Winn, in *Exoplanets*, ed. by S. Seager (Tucson, AZ: University of Arizona Press, 2011), pp. 55–77
J.N. Winn, R.W. Noyes, M.J. Holman et al., ApJ **631**, 1215 (2005)
J.N. Winn, J.A. Johnson, S. Albrecht et al., ApJ **703**, L99 (W09) (2009)

Chapter 5
Spin–Orbit Misalignments of Kepler-13Ab and HAT-P-7b from Gravity-Darkened Transit Light Curves

Abstract In Sect. 3.4, we discussed the importance of individual obliquity measurements for long-period planets around hot stars. Such measurements may be possible by applying the gravity-darkening method (Sect. 2.3.2) to existing data of transiting systems obtained by the *Kepler* space telescope. The methodology, however, is not fully established, given the discrepancy between this method and the spectroscopic one recently reported for the hot Jupiter system Kepler-13A. In this chapter, we discuss the origin of the discrepancy and present a possible solution. In addition, we show that the solution can be tested by future follow-up observations, on the basis of dynamical modeling of transit variations observed in this system. The revised methodology is then applied for the first time to the HAT-P-7 system, providing a useful cross-check between the gravity-darkening result and the measurement made in Chap. 4. The results presented in this chapter clarify the validity and limitation of the gravity-darkening method, and also demonstrate the potential of space-based photometry data to characterize exoplanets and their host stars.

Keywords Gravity darkening · Stellar oblateness · Spin–orbit precession
Kepler-13 · HAT-P-7

5.1 Introduction

Stellar obliquity or the spin–orbit angle, ψ, the angle between the stellar spin axis and the orbital axis of its planet, serves as a unique probe of the dynamical history of planetary systems. Especially, its connection with the hot-Jupiter migration has been extensively studied, but the relationship between the observed samples and the migration process is not straightforward for various reasons (see Chap. 3 for more detail). First of all, the initial distribution of the stellar obliquity is not known. Some studies do suggest that the protoplanetary disk may have already been misaligned with the stellar equator due to the chaotic gas accretion (e.g., Bate et al. 2010; Fielding et al. 2015) or the magnetic star–planet interaction (e.g., Lai et al. 2011). In these cases, the spin–orbit misalignment is primordial, rather than due to the migration. Even after the disk dissipation or the completion of migration, stellar obliquity can evolve due to the gravitational perturbation from the companion (e.g., Storch et al. 2014; Li et al. 2014). As suggested by the observed correlation between the spin–orbit misalignment and stellar effective temperature (Winn et al. 2010; Albrecht et al. 2012, see also Sect. 2.4), spin–orbit angle may also be affected by the tidal star–planet interaction (e.g., Xue et al. 2014), whose mechanism is not well understood. To partially resolve these issues, it is beneficial to measure stellar obliquities for systems with various host-star and orbital properties. For instance, planets on distant orbits or around hot/young stars are valuable targets because we expect that tides have not significantly affected the primordial spin–orbit configuration.

This chapter focuses on a relatively new method for the spin–orbit angle determination in transiting systems, which utilizes the gravity darkening of the host star owing to its rapid rotation (Barnes 2009, see also Sect. 2.3.2). Stellar rotation makes the effective surface gravity at the stellar equator smaller than that at the pole by a fractional order of $\gamma \equiv \Omega_\star^2 R_\star^3 / 2GM_\star \sim (P_{\rm br}/P_{\rm rot})^2$, where $\Omega_\star$, $R_\star$, $M_\star$, $P_{\rm br}$, and $P_{\rm rot}$ are angular rotation frequency, radius, mass, break-up rotation period, and rotation period of the star, respectively. According to von Zeipel's theorem (von Zeipel 1924), this results in the inhomogeneity of the stellar surface brightness through the relation $T_{\rm eff} \propto g_{\rm eff}^\beta$. Here, $T_{\rm eff}$ and $g_{\rm eff}$ are the effective temperature and surface gravity at each point on the stellar surface, and gravity-darkening exponent β characterizes the strength of the gravity darkening, which is theoretically 0.25 for a barotropic star with a radiative envelope. When a planet transits a stellar disk with such an inhomogeneous and generally non-axisymmetric brightness distribution, an anomaly of $\mathcal{O}(\gamma\delta)$ appears in the light curve, where δ is the transit depth. Since the shape of the anomaly depends on the position of the stellar pole relative to the planetary orbit, the stellar spin obliquity ψ can be estimated with the light-curve model taking into account the effect of gravity darkening.

Indeed, this "gravity-darkening method" has many unique aspects. So far, it is the only known method that is sensitive to *both* components of ψ, the sky-projected spin–orbit angle λ and stellar inclination $i_\star$ (cf. Eq. (5.2) and Fig. 2.1). Moreover, obliquity analysis is possible essentially with the photometric data alone, and its application is not necessarily limited to short-period planets, as far as the transit is observed with sufficient signal-to-noise ratio (Zhou and Huang 2013). It is also interesting to note that the method is (only) applicable to fast-rotating (i.e., young or hot) stars, for which anomalies of larger amplitudes result. Since rapid rotators are not suitable for the precise spectroscopic velocimetry because of their broad spectral lines, this method is complementary to the conventional spin–orbit angle measurement using the Rossiter-McLaughlin (RM) effect. All these properties make the method suitable for sampling stars for which tidal effect is not so significant that the primordial information is expected to be well preserved in the current spin–orbit configuration.

Although the gravity-darkening method is valuable in many aspects, the procedure for obtaining ψ may not be fully established. In a representative example of its application, Kepler-13A, the constraint from the gravity-darkening method (Barnes et al. 2011, hereafter B11) is known to be in disagreement with the later spectroscopic measurement of λ with the Doppler tomography (Johnson et al. 2014, see also Sect. 2.2). In addition, inconsistent results arise even within the gravity-darkening analyses, depending on the choice of the limb-darkening coefficients or β (Zhou and Huang 2013; Ahlers et al. 2014). For these reasons, it is worth revisiting the reliability and limitation of this method more carefully, in order for this unique method to be applied to more systems in future and provide credible results.

In this chapter, we reanalyze a well-known example of the gravity-darkened transit of Kepler-13Ab, with more data than used in the previous analysis by B11. We investigate the systematic effects in the spin–orbit angle determination, and propose a joint solution that may solve the discrepancy with the Doppler tomography measurement (Sect. 5.3). We will also show that the spin–orbit precession observed in this system can be used to test the validity of our solution, as well as to determine the stellar quadrupole moment J_2 (Sect. 5.4).

In addition, we apply the gravity-darkening method for the first time to an F-type dwarf star, HAT-P-7, where the anomaly in the transit light curve has been reported in several studies (e.g., Esteves et al. 2013, 2015; Van Eylen et al. 2013, see also Chap. 4). While the RM measurements (Winn et al. 2009a; Narita et al. 2009; Albrecht et al. 2012) have established that $\lambda > 90°$, suggesting a retrograde orbit, the following asteroseismic inferences (Chap. 4; Lund et al. 2014) have revealed that a pole-on orbit is actually favored. In Sect. 5.5, we show that a similar conclusion is also obtained from the gravity-darkening method and discuss the consistency of our result with other constraints on the host-star properties.

5.2 Method

5.2.1 Model

We basically follow Barnes (2009) in modeling the gravity-darkened transit light curve. The model includes the following 14 parameters, which are listed as "fitting parameters" in Tables 5.1 and 5.3:

1. mean stellar density, $\rho_\star = 3M_\star/4\pi R_\star^3$, which corresponds to the semi-major axis scaled by the stellar equatorial radius, $a/R_\star$[1]
2. limb-darkening coefficient for the quadratic law, $c_1 = u_1 + u_2$,
3. limb-darkening coefficient for the quadratic law, $c_2 = u_1 - u_2$,
4. time of the inferior conjunction (where the planet is closest to the observer), t_c,
5. orbital period, P,
6. cosine of orbital inclination, $\cos i_{\text{orb}}$,
7. planetary radius normalized to the stellar equatorial radius, $R_p/R_\star$
8. normalization of the out-of-transit flux, F_0
9. stellar mass, $M_\star$,

---

[1] In this chapter, $R_\star$ denotes the equatorial radius of the star.

10. stellar rotation frequency, $f_{\rm rot}$
11. stellar effective temperature at the pole, $T_{\star,\rm pole}$
12. gravity-darkening exponent, β,
13. stellar inclination, $i_\star$
14. sky-projected spin–orbit angle, λ.

See also Appendix B.4 for the justification of these choices.

The first eight parameters are common with the light-curve model without gravity darkening. We assume circular orbits for the two targets because the orbital eccentricities are constrained to be very small, if any, from the occultation light curves (Chap. 4, Shporer et al. 2014).

Following the gravity-darkened model by Barnes (2009), the shape of the star is approximated by the spheroid with the oblateness $\gamma = \Omega_\star^2 R_\star^3 / 2GM_\star = 3\pi f_{\rm rot}^2 / 2G\rho_\star$. The surface brightness at each point is modeled as the blackbody emission of the temperature $T_\star = T_{\star,\rm pole} \left(g_{\rm eff}/g_{\rm eff,pole} \right)^\beta$, where $g_{\rm eff}/g_{\rm eff,pole}$ is the effective surface gravity normalized by its value at the stellar pole. The surface gravity at point r on the stellar surface is calculated by $\boldsymbol{g}_{\rm eff} = -GM_\star r^{-2} \hat{\boldsymbol{r}} + 4\pi^2 f_{\rm rot}^2 r_\perp \hat{\boldsymbol{r}}_\perp$. Here r and $\hat{\boldsymbol{r}}$ are the norm and unit vector of the radius vector $\boldsymbol{r}$, respectively. Similarly, $r_\perp$ and $\hat{\boldsymbol{r}}_\perp$ are those of $\boldsymbol{r}_\perp$, the projection of $\boldsymbol{r}$ onto the stellar equatorial plane. The Planck function $B_\lambda(T_\star)$ at each point is convolved with the "high-resolution" *Kepler* response function[2] using the table of the wavelength- and temperature-dependent factor calculated prior to the fitting. The convolved flux is then multiplied by the limb-darkening function

$$I(\mu) = 1 - u_1(1-\mu) - u_2(1-\mu)^2, \quad (5.1)$$

with μ being the cosine of the angle between $-\boldsymbol{g}_{\rm eff}$ and our line of sight,[3] and integrated over the visible surface of the star to give the total flux. We fix $T_{\star,\rm pole}$ at the observed effective temperature assuming that the difference between $T_{\star,\rm pole}$ and the disk-integrated effective temperature is small. Note that the gravity-darkened transit light curve gives $\rho_\star$ alone and cannot constrain $M_\star$ and $R_\star$ separately, as is the case for the transit without gravity darkening (cf. Appendix B.4).

The configuration of the planetary orbit and stellar spin is specified by three angles, $i_{\rm orb}$, $i_\star$, and λ, which are defined in Fig. 2.1 (see also Fig. 4.1). The orbital and stellar inclinations, $i_{\rm orb}$ and $i_\star$, are measured from the line of sight and defined to be in the range $[0, \pi]$. The sky-projected spin–orbit angle, λ, is the angle between the sky-projected stellar spin and planetary orbital axes. It is measured from the former to the latter counterclockwise in the sky plane, and is in the range $[0, 2\pi]$. With these definitions, the true spin–orbit angle, or the stellar obliquity, ψ, is given by Eq. (4.1):

$$\cos \psi = \cos i_\star \cos i_{\rm orb} + \sin i_\star \sin i_{\rm orb} \cos \lambda. \quad (5.2)$$

---

[2] http://keplergo.arc.nasa.gov/CalibrationResponse.shtml.
[3] Although this vector $-\boldsymbol{g}_{\rm eff}$ is not exactly parallel to the surface normal of the spheroid we assume, the difference is $\mathcal{O}(\gamma^2)$ and thus negligible.

5.2 Method

Throughout the chapter, we restrict $i_\star$ to be in the range $[0, \pi/2]$ making use of the intrinsic symmetry with respect to the sky plane. We do not lose any physical information of the system with this choice because any of the relative star–planet configurations with $i_\star$ in $[\pi/2, \pi]$ is the same as one of those with $i_\star$ in $[0, \pi/2]$. In other words, the configurations $(i_\star, i_{\mathrm{orb}}, \lambda)$ and $(\pi - i_\star, \pi - i_{\mathrm{orb}}, -\lambda)$ are equivalent. This transformation corresponds to looking at the system from the other side of the plane of the sky.

In the following, we also adopt the constraint on the stellar line-of-sight rotational velocity $v \sin i_\star$ from spectroscopy, which is related to the above model parameters by

$$v \sin i_\star = 2\pi f_{\mathrm{rot}} \left(\frac{3 M_\star}{4\pi \rho_\star} \right)^{1/3} \sin i_\star. \tag{5.3}$$

This, in principle, allows us to break the degeneracy between $M_\star$ and $R_\star$, enabling the determination of the absolute dimension of the system. Nevertheless, the constraint on $M_\star$ is usually weak due to the $M_\star^{1/3}$ dependence, and so we fix $M_\star$ at the observed value.

5.2.2 Data Processing

We detrend and normalize the transit light curves of each target along with the consistent determination of the transit times and transit parameters. We first normalize the light curve of each quarter using its median, and then iterate the following two steps (typically 10–20 times) until the resulting transit times t_c and transit parameters converge:

1. Light curve around each transit (± 0.2 days for Kepler-13A and ± 0.15 days for HAT-P-7) is modeled as the product of a quadratic polynomial[4] $a_0 + a_1(t - t_c) + a_2(t - t_c)^2$ (t: time) and the analytic transit light-curve model by Mandel and Agol (2002). We use the Levenberg-Markwardt (LM) method (Markwardt 2009) to fit a_0, a_1, a_2, and t_c iteratively removing 5σ outliers, while the other parameters are fixed. The filtered data are then divided by the best-fit polynomial to give a normalized and detrended transit light curve. We discard the transits with data gaps of more than 50%.
2. Using the set of t_c obtained in the first step, we calculate the mean orbital period P and transit epoch t_0 by linear fit and use them to phase-fold the normalized and detrended transits. The phase-folded light curve is averaged into one-minute bin and then fitted with the Mandel and Agol (2002) model using an LM algorithm. We fit c_1, c_2, $\rho_\star$, $\cos i_{\mathrm{orb}}$, $R_{\mathrm{p}}/R_\star$, and F_0, whose best-fit values are used in the

---

[4] Use of the quadratic polynomial helps the better removal of flux variation not due to the transit, i.e., planetary light, ellipsoidal variation, and Doppler beaming.

step 1 of the next iteration. In this step, the orbital period P is fixed to be the value obtained from the linear fit and the central time of the phase-folded transit is fixed to be zero.

In the following analysis, we use the one-minute binned, phase-folded light curve obtained in the second step of the final iteration. For each bin, the flux value is given by its mean and the error is estimated as the standard deviation within the bin divided by the square root of the number of data points.

5.2.3 Fitting Procedure

In fitting the observed light curves, the likelihood $\mathcal{L}$ of the model is computed by $\mathcal{L} \propto \exp(-\chi^2/2)$, where

$$\chi^2 = \sum_i \left(\frac{f_i - f_{\text{model},i}}{\sigma_i} \right)^2 + \sum_j \left(\frac{p_j - p_{\text{model},j}}{\delta p_j} \right)^2. \quad (5.4)$$

In the first term, f_i, $f_{\text{model},i}$, and σ_i are the observed value, modeled value, and error of the ith flux data. The second term is introduced to take into account the constraints from other observations on some (functions) of the model parameters p_j. In the following analysis, p is read to be $v \sin i_\star$ and, in some cases, λ.[5] For each p_j, we assume a Gaussian constraint of the form $p_j \pm \delta p_j$ and the value obtained from the model is denoted by $p_{\text{model},j}$.

The maximum likelihood solution is found by minimizing Eq. (5.4) with the LM method using the cmpfit package (Markwardt 2009). Since the complex dependence of χ^2 on $i_\star$ and λ is expected, we repeat the fitting procedure from the initial $i_\star$ in [0, 90°] and λ in [−180°, 180°] at 10° intervals. Initial values of the other parameters are chosen close to the best-fit values obtained from the model without gravity darkening. We also try both positive and negative $\cos i_{\text{orb}}$ as an initial value to search the whole domain of i_{orb}, which is now [0°, 180°].

5.3 Transit Analysis of Kepler-13Ab

In this section, we report the analysis of the gravity-darkened transit of Kepler-13Ab. We first analyze the whole available data using the same stellar parameters as in B11 to test the validity of our method (Sect. 5.3.1). Motivated by the recently reported disagreement with λ from the Doppler tomography, we also investigate the possible systematics in the spin–orbit determination arising from the choice of

---

[5] Only in Sect. 5.4.1, $\rho_\star$, c_1, c_2, $R_p/R_\star$, and f_{rot} are also included.

5.3 Transit Analysis of Kepler-13Ab

stellar parameters. We show that the discrepancy can be solved by adjusting the value of c_2 and present a joint solution that is compatible with all of the observations made so far.

5.3.1 Reproducing the Results by B11

In this subsection, we analyze the short-cadence (SC), Pre-search Data Conditioned Simple Aperture Photometry (PDCSAP) fluxes from Q2, 3, and 7–17. Note that only the Q2 data were available when B11 analyzed this system. Given the clear transit duration variation (TDV) reported by Szabó et al. (2012, 2014), we separately analyze the transits from each quarter, rather than folding all the available data. Since we do not detect significant temporal variations in the parameters other than $\cos i_{\rm orb}$ (see Sect. 5.4), we report the mean and standard deviation of the best-fit values from the above 13 quarters for each parameter.

First, we use the same stellar parameters as in B11 and obtain the results in the second column of Table 5.1. Namely, we subtract a constant value $F_c = 0.45$ from the normalized flux to remove the flux contamination from the companion star, and impose the constraint $v \sin i_\star = 65 \pm 10 \,\mathrm{km\,s^{-1}}$ based on Szabó et al. (2011). We fix $M_\star = 1.83 M_\odot$ and $T_{\star,\rm pole} = 8848$ K from Borucki et al. (2011), and $c_2 = 0$. In Fig. 5.1, the best-fit model is overplotted with the data for Q2, which is to be compared with Fig. 5.2 of B11.

Basically, we find a very good agreement with the result by B11 using about 12 times more data. Although the values of $\cos i_{\rm orb}$, $i_\star$, and λ we report here appear different from those in B11, that is simply because we choose $i_\star$ to be in the range $[0, \pi/2]$. This is physically the same configuration as theirs and corresponds to the top-left situation in Fig. 5.3 of B11. That is, *λ in our solutions with $\cos i_{\rm orb} < 0$ should be read as $-\lambda$ in the conventional definition, because λ is usually defined for the orbit with $\cos i_{\rm orb} > 0$* (see also the discussion after Eq. 5.2).

In addition to the solution in Table 5.1, we also find a retrograde solution with $\lambda > 90°$ as noted in B11. Here we do not discuss this solution, however, because the Doppler tomography observation has already excluded the retrograde orbit with high significance (Johnson et al. 2014).

5.3.2 Systematics Due to Stellar Parameters

Although we find consistent values of λ and $i_\star$ as obtained by B11, those of λ significantly differ from $\lambda = 58°\!.6 \pm 2°\!.0$, the value obtained from the Doppler tomography (Johnson et al. 2014). Motivated by this discrepancy, we investigate the possible origins of systematics in the spin–orbit angle determination with gravity darkening in this subsection.

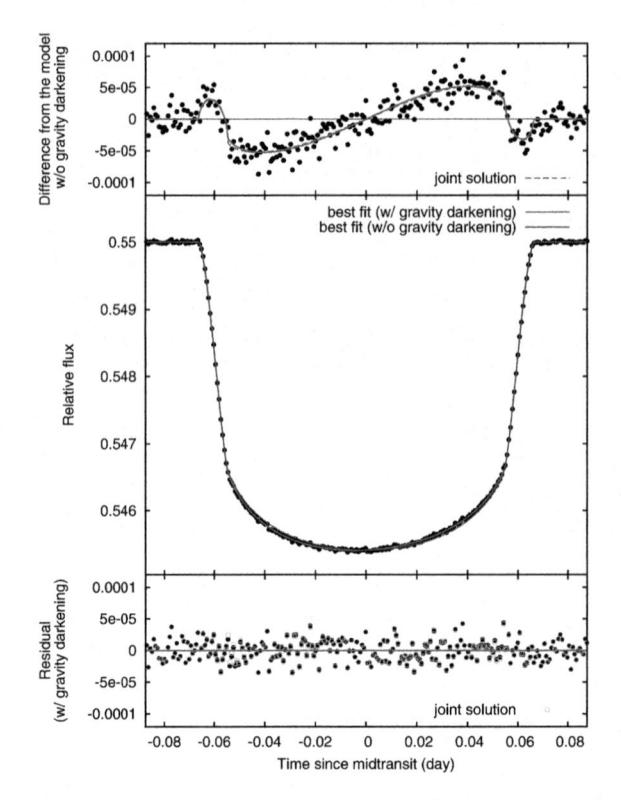

Fig. 5.1 Fitting the gravity-darkened model to the Q2 transit of Kepler-13Ab. (Middle) Black dots are the phase-folded and binned fluxes from Q2. The thick red line shows our best-fit gravity-darkened model, while the thin blue line is the best-fit model without gravity darkening. (Bottom) Black dots are the residual of the best-fit gravity-darkened model. Gray open circles are those for the joint solution, where c_2 is fitted with the constraint $\lambda = 58°.6 \pm 2°.0$ from the Doppler tomography. (Top) Black dots are the residuals of the best-fit model without gravity darkening. Thick red line is the difference between the best-fit model with gravity darkening and that without gravity darkening. Dashed red line shows the same result for the joint solution. The difference between the two gravity-darkened solutions is only barely visible just after the ingress and before the egress

First, we examine the systematics due to the choice of $M_\star$, $v \sin i_\star$, $T_{\star,\mathrm{pole}}$, and F_c, which are the stellar properties not derived from the light curve modeling.[6] We perform the same analysis as in Sect. 5.3.1, but adopting the following parameters from the most recent photometric and spectroscopic study by Shporer et al. (2014, hereafter S14): $v \sin i_\star = 78 \pm 15 \,\mathrm{km\,s^{-1}}$, $M_\star = 1.72 M_\odot$, $T_{\star,\mathrm{pole}} = 7650\,\mathrm{K}$, and $F_c = 0.47726$. The corresponding results are shown in the third column of Table 5.1. We find that $i_\star$ and λ can differ by as large as $10°$ due to the choice of the above parameters, but the difference is not so large as to explain the disagree-

---

[6] We do not examine the dependence on β here because B11 have already shown that a different choice of $\beta = 0.19$, suggested by the interferometric observation of Altair (Monnier et al. 2007), does not change the result significantly.

5.3 Transit Analysis of Kepler-13Ab

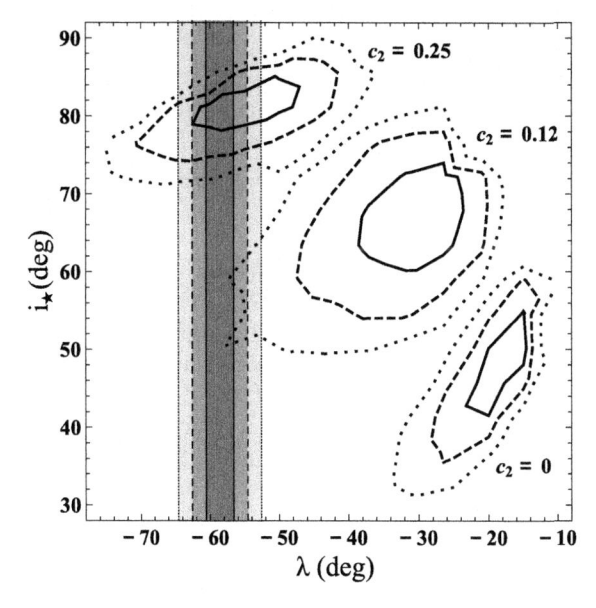

Fig. 5.2 Constraints on $(\lambda, i_\star)$ from the gravity-darkened transit of Kepler-13Ab for the different choices of c_2. In this illustration, data from Q2 are used and stellar parameters from B11 are adopted. The solid, dashed, and dotted contours respectively show 1σ, 2σ, and 3σ confidence regions for $(\lambda, i_\star)$ obtained from 200000 Markov Chain Monte Carlo (MCMC) samples for three fixed values of c_2 (0, 0.12, and 0.25). The shaded areas bounded by the vertical solid, dashed, and dotted lines respectively denote 1σ, 2σ, and 3σ confidence regions for λ obtained from the Doppler tomography (Johnson et al. 2014). The sign of λ is opposite to their quoted value because we are now dealing with the solution with $\cos i_{\rm orb} < 0$ (i.e., $\pi/2 < i_{\rm orb} < \pi$); see also the discussion in the third paragraph of Sect. 5.3.1

ment with the Doppler tomography. The main difference from the B11 case with this new set of parameters is the different constraint on $f_{\rm rot} \sin i_\star$, which is proportional to the combination $(\rho_\star/M_\star)^{1/3} v \sin i_\star$ (cf. Eq. 5.3). With smaller $M_\star$ and larger $v \sin i_\star$, the stellar rotation rate slightly higher than the B11 case is favored. We find that the difference in $T_{\star,\rm pole}$ is less important compared to the above effect. We also find that larger F_c yields larger $R_p/R_\star$, which makes the impact parameter or $|\cos i_{\rm orb}|$ smaller to give the same ingress/egress duration.

Next, we allow $c_2 = u_1 - u_2$ to be free, and find that the resulting spin–orbit angle is very sensitive to this parameter. When c_2 is floated, the constraints on $i_\star$ and λ become much weaker than the $c_2 = 0$ case, as shown in the fourth and fifth columns of Table 5.1. The strong dependence on c_2 is illustrated in Fig. 5.2, which shows that λ and $i_\star$ vary by several tens of degrees depending on c_2; see also the joint MCMC posterior distribution in Fig. C.5 for the same data. In fact, the result indicates that the gravity-darkened light curve is actually compatible with the Doppler tomography solution if we choose $c_2 \sim 0.25$; such a solution will be discussed in Sect. 5.3.3.

5.3.3 Joint Solution

In Sect. 5.3.2, we found that the gravity-darkened light curve is compatible with the value of λ estimated from the Doppler tomography if $c_2 \sim 0.25$. Thus we repeat the analysis treating c_2 as a free parameter for both stellar parameters by B11 and S14, but this time imposing additional constraint $\lambda = 58°\!.6 \pm 2°\!.0$ from the Doppler tomography. The results are summarized in the last two columns in Table 5.1, and the joint posterior distribution from the MCMC fitting to the Q2 data is shown in Fig. C.6 for the B11 stellar parameters. The resulting value of $i_\star = 81° \pm 5°$ indicates that the star is close to equator-on, and $\psi = 60° \pm 2°$ is slightly larger than the previous estimate. In terms of $\chi^2_{\min}$, these solutions equally well reproduce the transit anomaly as the solutions discussed so far, and still they are consistent with the Doppler tomography result. Moreover, we obtain a slightly longer $P_{\rm rot}$, which better agrees with $P_{\rm rot} = 25.43 \pm 0.05$ h estimated by Szabó et al. (2012, 2014) than the solution with the gravity darkening alone. For these reasons, the joint solution is most favored from the current observations.

We note, however, that the likelihood for the joint solution is not so high as to statistically justify the introduction of the additional free parameter c_2. Furthermore, the plausibility of the value of c_2 in our joint solution is theoretically unclear. We obtain the theoretical values of $c_{1,\rm th} \simeq 0.6$ and $c_{2,\rm th} \simeq 0.0$ from the table of Sing (2010) if we adopt the effective temperature and surface gravity by S14. Hence the value of c_2 from our joint solution is discrepant from $c_{2,\rm th}$. Nevertheless, it is also true that theoretical values often disagree with the observed ones (e.g., Southworth 2008); in fact, c_1 in the light-curve solution with $c_2 = 0$ is also different from $c_{1,\rm th}$. Therefore, we do not consider the possible deviations from the theoretical values crucial, and regard it as an open question.[7] An alternative approach to independently assess the validity of our solution is discussed in the next section.

5.4 Spin–Orbit Precession in the Kepler-13A System

The shape of Kepler-13Ab's transit is known to exhibit a long-term variation, which is likely due to the spin–orbit precession induced by the quadrupole moment of the rapidly rotating host star (Szabó et al. 2012, 2014). Indeed, we find the monotonic decrease in $|\cos i_{\rm orb}|$ from the quarter-by-quarter analysis in Sect. 5.3; the constant-value model is rejected at the p-value of 0.5% for this parameter using a simple χ^2 test. On the other hand, the other model parameters are found to be consistent with the constant value using the same criterion. Therefore, our analysis confirms that the observed TDVs are actually due to the variation in $\cos i_{\rm orb}$,[8] further supporting

---

[7]For reference, we find $c_2 = 0.1 - 0.2$ if we adopt the model without gravity darkening (Mandel and Agol 2002), which suggests that the choice of $c_2 = 0$ is not indispensable.

[8]Note that, in Szabó et al. (2012), the degeneracy between $a/R_\star$ (or $\rho_\star$) and $\cos i_{\rm orb}$ was not solved.

5.4 Spin–Orbit Precession in the Kepler-13A System

the precession scenario with the more realistic model of the asymmetric transit light curve.

In this section, we further examine this scenario with the gravity-darkened transit model. Unlike the above previous studies (Szabó et al. 2012, 2014) that focused on $i_{\rm orb}$, the gravity-darkened model allows us to additionally study the (non-)variations in the other two angles, λ and $i_\star$, which should also be induced if the system is precessing.[9] By fitting the analytic precession model to the time series of $\cos i_{\rm orb}$, λ, and $i_\star$ obtained from the light curves, we constrain the stellar quadrupole moment J_2 and its moment of inertia coefficient $\mathbb{C}$. On the basis of these constraints, we predict the future evolution of the system configuration and argue that the follow-up observations of such a long-term modulation can distinguish the light-curve and joint solutions discussed in Sect. 5.3. In the following, we mainly discuss the results obtained with the stellar parameters from S14, though the conclusions remain the same for the B11 parameters.

5.4.1 Model Parameters from Each Transit

To examine the temporal variations in $\cos i_{\rm orb}$, $i_\star$, and λ, we fit individual transit light curves, rather than the phase-folded ones, for these parameters. We use the same two models ("light-curve solution" with $c_2 = 0$ and "joint solution" with c_2 fitted) as discussed in Sect. 5.3. In order not to underestimate the errors in the three angles, we fit all the other model parameters, $\rho_\star$, c_1, c_2 (for the joint model), t_c, $R_{\rm p}/R_\star$, $f_{\rm rot}$, and F_0 as well, which should not vary temporally in our model. Using the best values in Table 5.1, we impose the constraints on these parameters except for t_c and F_0, through the second term of Eq. (5.4). In fitting much noisier individual transits, this prescription assures that the parameters converge to the values consistent with those from the phase-folded light curves, while preserving their differences from transit to transit. We also discard the transits for which the fit does not converge due to the data gaps and/or short brightening features sometimes found in the light curves. The resulting sequences of the transit parameters are plotted in Fig. 5.3.

As mentioned above, we again find the clear linear trend in $\cos i_{\rm orb}$ from individual transits. We fit the linear model to the time series of $\cos i_{\rm orb}$ using a Markov Chain Monte Carlo (MCMC) algorithm and obtain the rates of change in the upper part of Table 5.2. Here we only report the slopes for absolute values of $\cos i_{\rm orb}$ because its actual sign depends on the sign of $\cos i_\star$, which can never be determined with the current observations (we arbitrarily choose $\cos i_\star > 0$ in this chapter, as discussed after Eq. 5.2). Comparing the light-curve solution and joint solution, we find that the rate of $|\cos i_{\rm orb}|$ change is insensitive to λ or c_2 because $|\cos i_{\rm orb}|$ is mainly

---

[9]If either of the angular momenta of the stellar spin or the orbital motion dominates, $i_{\rm orb}$ or $i_\star$ is almost constant. In the Kepler-13A system, the two angular momenta have comparable magnitudes and so all three angles modulate due to the precession. A similar case, the PTFO 8-8695 system, has been studied by Barnes et al. (2013) and Kamiaka et al. (2015).

Table 5.1 Results for the transit of Kepler-13Ab

	Light-curve solution ($c_2=0$)		Light-curve solution (c_2 fitted)		Joint solution (c_2 fitted)	
Ref. for F_c, $v\sin i_\star$, $M_\star$, $T_{\star,\text{pole}}$	B11	S14	B11	S14	B11	S14
(Assumed flux contamination)						
F_c	0.45	0.47726	0.45	0.47726	0.45	0.47726
(Constraints)						
$v\sin i_\star$ (km s^{-1})	65 ± 10	78 ± 15	65 ± 10	78 ± 15	65 ± 10	78 ± 15
λ (deg)	…	…	…	…	58.6 ± 2.0[a]	58.6 ± 2.0[a]
(Fitting parameters)						
$M_\star$ ($M_\odot$)	1.83 (fixed)	1.72 (fixed)	1.83 (fixed)	1.72 (fixed)	1.83 (fixed)	1.72 (fixed)
$T_{\star,\text{pole}}$ (K)	8848 (fixed)	7650 (fixed)	8848 (fixed)	7650 (fixed)	8848 (fixed)	7650 (fixed)
$\rho_\star$ (g cm^{-3})	0.533 ± 0.005	0.550 ± 0.005	0.530 ± 0.005	0.547 ± 0.006	0.525 ± 0.005	0.538 ± 0.006
c_1	0.496 ± 0.008	0.493 ± 0.008	0.50 ± 0.04	0.51 ± 0.02	0.523 ± 0.005	0.528 ± 0.006
c_2	0 (fixed)	0 (fixed)	0.02 ± 0.28	0.12 ± 0.13	0.20 ± 0.02	0.26 ± 0.04
t_c (10^{-5} day)[b]	−3 ± 1	−3 ± 1	−8 ± 7	−10 ± 7	−9 ± 4	−11 ± 5
P (day)			1.763587 ± 0.000002			
$\cos i_{\text{orb}}$	−0.066 ± 0.004	−0.057 ± 0.004	−0.066 ± 0.003	−0.055 ± 0.004	−0.064 ± 0.004	−0.054 ± 0.004
$R_p/R_\star$	0.0845 ± 0.0002	0.0864 ± 0.0002	0.0845 ± 0.0003	0.0865 ± 0.0002	0.0846 ± 0.0002	0.0864 ± 0.0003
F_0			0.550000 ± 0.000002(B11)/0.522740 ± 0.000002(S14)			
f_{rot} (μHz)	12.9 ± 0.4	14.5 ± 0.6	12.8 ± 4.2	12.9 ± 1.8	10.2 ± 0.6	11.6 ± 1.0
$i_\star$ (deg)	47 ± 3	56 ± 3	60 ± 20	71 ± 16	73 ± 5	81 ± 5
λ (deg)	−20.3 ± 1.3	−13.9 ± 1.3	−33 ± 13	−30 ± 12	−58.4 ± 2.0[c]	−58.5 ± 2.0[c]
β	0.25 (fixed)	0.25 (fixed)	0.25 (fixed)	0.25 (fixed)	0.25 (fixed)	0.25 (fixed)

(continued)

5.4 Spin–Orbit Precession in the Kepler-13A System

Table 5.1 (continued)

Ref. for F_c, $v \sin i_\star$, $M_\star$, $T_{\star,\text{pole}}$	Light-curve solution ($c_2 = 0$) B11	S14	Light-curve solution (c_2 fitted) B11	S14	Joint solution (c_2 fitted) B11	S14
(Derived parameters)						
P_rot (hr)	21.5 ± 0.7	19.1 ± 0.8	23 ± 5	22 ± 3	27 ± 2	24 ± 2
ψ (deg)	50 ± 3	40 ± 3	52 ± 9	42 ± 6	61 ± 2^c	60 ± 2^c
Impact parameter	0.29 ± 0.02	0.26 ± 0.02	0.29 ± 0.01	0.25 ± 0.02	0.28 ± 0.02	0.24 ± 0.02
Stellar oblateness	0.022 ± 0.001	0.027 ± 0.002	0.02 ± 0.02	0.022 ± 0.006	0.014 ± 0.002	0.018 ± 0.003
$\chi^2_\text{min}/\text{dof}$	250/241	249/241	247/240	245/240	248/241	246/241

Note The quoted best-fit values and uncertainties are averages and standard deviations of the best-fit values obtained from 13 quarters analyzed here. The value of χ^2_min is also the average of the minimum χ^2 among quarters

[a] This should be read as $-58°.6 \pm 2°.0$ for the solution with $\cos i_\text{orb} < 0$ discussed in this table

[b] Measured from the transit epoch $t_0(\text{BJD} - 2454833) = 120.566 \pm 0.001$ obtained with the transit model without gravity darkening

[c] For λ in the joint solutions, we quote the uncertainty in the constraint from the Doppler tomography. This is because the value of λ is completely determined by this constraint and its standard deviation (several $0.1°$) is not a good measure of the actual uncertainty. Accordingly, the quoted uncertainty in ψ is also increased by taking a quadratic sum of its standard deviation and the additional scatter coming from the uncertainty of $2°$ in λ

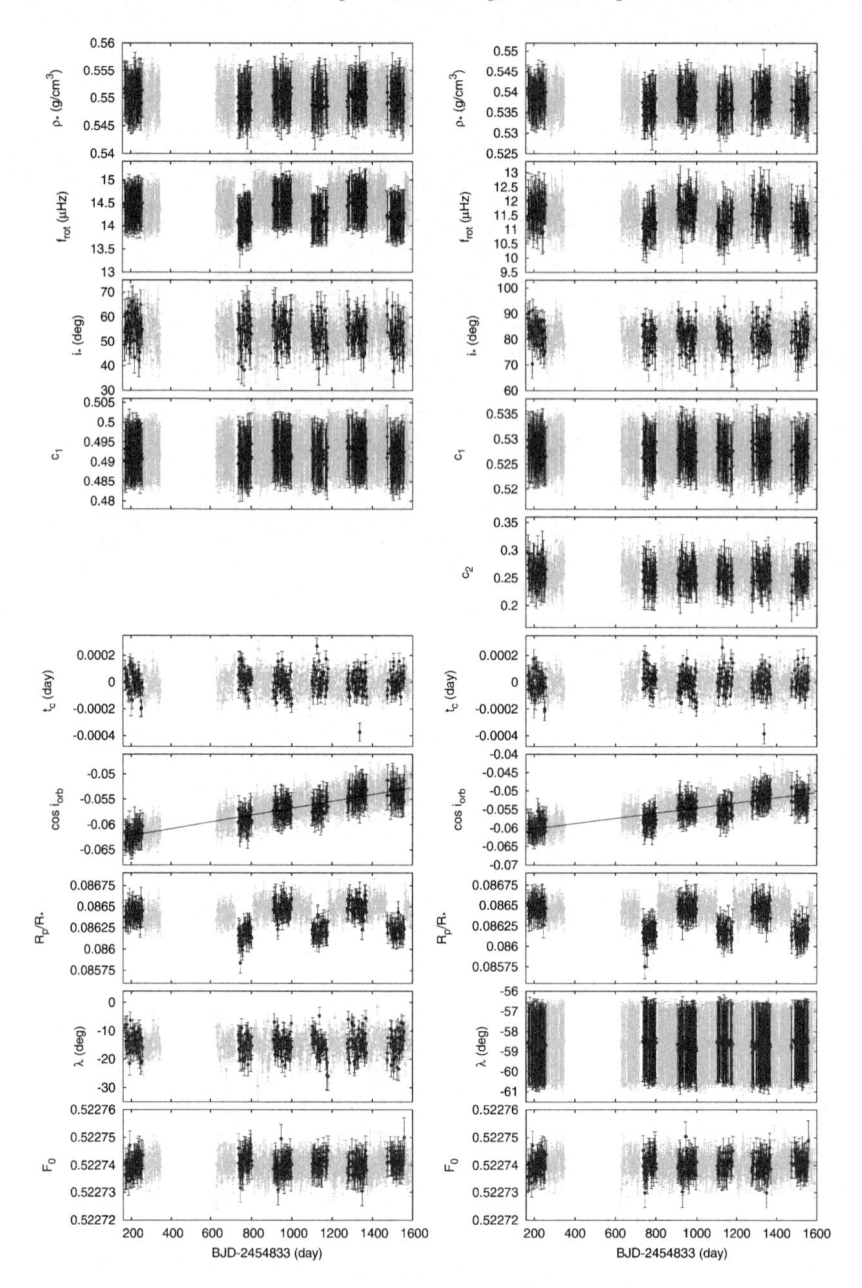

Fig. 5.3 Best-fit model parameters from each transit. The left panels are the results for the light-curve solution with $c_2 = 0$, while the right ones are for the joint solution. Errors are from the outputs of the `cmpfit` package. Parameters from even quarters (2, 8, 10, 12, 14, and 16) are shown in black, while those from odd quarters (3, 7, 9, 11, 13, 15, and 17) are in gray. For the times of inferior conjunctions, t_c, the residuals of the linear fit (i.e., TTVs) are plotted for clarity. Solid lines in $\cos i_{\rm orb}$ panels are the best-fit linear models

determined from the transit duration. With $a/R_\star$ calculated from $\rho_\star$, our value for $d|\cos i_{\rm orb}|/dt$ is found to be consistent with $db/dt = (-4.4 \pm 1.2) \times 10^{-5}$ day^{-1} by Szabó et al. (2012), but our constraint is several times better.

Figure 5.3 also shows the abrupt systematic changes in $R_{\rm p}/R_\star$. These changes occur exactly in phase with the border of different quarters indicated with different colors (black and gray). For this reason, they are unlikely to be of physical origin, but are probably due to the seasonal transit depth variations similar to those reported by Van Eylen et al. (2013) for HAT-P-7. In addition, some of the parameters (most notably $\rho_\star$ and $f_{\rm rot}$) show the long-term modulation of the period $\sim$400 days. Origins of these systematics are beyond the scope of this chapter, and they are just treated as the additional scatter in the data.

5.4.2 Fit to the Observed Angles and Future Prediction

Among the observed time series of transit parameters in Fig. 5.3, those of $\cos i_{\rm orb}$, λ, and $i_\star$ are fitted using an MCMC algorithm to observationally constrain J_2 and $\mathbb{C}$. We utilize the same analytic precession model as in Barnes et al. (2013), which constitutes an analytic solution of the secular equations of motion derived by Boue and Laskar (2009). In this model, the orbital and spin angular momenta precess around the total angular momentum at the same angular rate given by

$$\dot{\Omega} = \dot{\Omega}_{\rm p} \sqrt{\left(\frac{L}{S} + \cos\psi\right)^2 + \sin^2\psi}, \qquad (5.5)$$

where $\dot{\Omega}_{\rm p}$ is the precession rate of the orbital angular momentum around the stellar spin, and explicitly given by

$$\dot{\Omega}_{\rm p} = -\frac{3}{2} J_2 \frac{2\pi}{P} \cos\psi \left(\frac{R_\star}{a}\right)^2 \qquad (5.6)$$

with J_2 being the stellar quadrupole moment. In the Kepler-13A system, the spin angular momentum, S, is comparable to the orbital one, L, owing to the small semimajor axis and rapid stellar rotation. As a consequence, $\dot{\Omega}$ also depends on the ratio of the two,

$$\frac{L}{S} = \frac{1}{\mathbb{C}} \frac{M_{\rm p}}{M_\star} \frac{1}{P f_{\rm rot}} \left(\frac{a}{R_\star}\right)^2, \qquad (5.7)$$

where $\mathbb{C}$ is the moment of inertia coefficient of the host star. Thus, the independent model parameters are $\rho_\star$, $f_{\rm rot}$, J_2, $\mathbb{C}$, P, $M_{\rm p}/M_\star$, and three angles $\cos i_{\rm orb}$, λ, $i_\star$ at some epoch (here taken to be BJD = 2454833 + 800). We do not relate J_2 to the other parameters like the stellar oblateness as done in Barnes et al. (2013).

To realistically evaluate the credible intervals of J_2 and $\mathbb{C}$ by marginalization, uncertainties in $\rho_\star$, f_{rot}, P, and $M_{\text{p}}/M_\star$ should also be taken into account. However, these parameters are not well determined from the data of $\cos i_{\text{orb}}$, λ, and $i_\star$. Thus, they are floated with the following Gaussian priors. The first three are assigned the same central values and widths as in Table 5.1. For the mass ratio, we take the mean and standard deviation of the results reported by S14, Esteves et al. (2015), and Faigler and Mazeh (2015), which come from the amplitudes of the ellipsoidal variation and Doppler beaming. We also impose the Gaussian prior on $\mathbb{C}$ centered on 0.0776 (the value for $n = 3$ polytrope by Szabó et al. 2012) and with the width of 0.02, which is chosen to enclose the solar value, 0.059.

The constraints from the MCMC fit are summarized in the middle and bottom parts of Table 5.2 and the best-fit models are plotted with the solid lines in Fig. 5.4.

Table 5.2 Results of the precession model fit to $\cos i_{\text{orb}}$, $i_\star$, and λ from each transit

	Light-curve solution ($c_2 = 0$)		Joint solution (c_2 fitted)	
Ref. for F_c, $v \sin i_\star$, $M_\star$, $T_{\star,\text{pole}}$	B11	S14	B11	S14
(Linear fit to $\cos i_{\text{orb}}$)				
$\lvert \cos i_{\text{orb}} \rvert$ [a]	0.0668 ± 0.0001	0.0581 ± 0.0001	0.0658 ± 0.0001	0.0560 ± 0.0002
$\frac{d\lvert \cos i_{\text{orb}}\rvert}{dt}$ (day^{-1})	$(-5.9 \pm 0.3) \times 10^{-6}$	$(-6.8 \pm 0.3) \times 10^{-6}$	$(-6.0 \pm 0.3) \times 10^{-6}$	$(-7.0 \pm 0.4) \times 10^{-6}$
(Precession model fit to $\cos i_{\text{orb}}$, $i_\star$, and λ)				
$\rho_\star$, f_{rot}, P	Same as Table 5.1 (priors = posteriors)			
$\cos i_{\text{orb}}$ [a]	-0.0668 ± 0.0001	-0.0581 ± 0.0001	-0.0658 ± 0.0001	-0.0560 ± 0.0002
$i_\star$ (deg)[a]	44.7 ± 0.3	54.2 ± 0.3	72.8 ± 0.3	81.8 ± 0.2
λ (deg)[a]	-20.1 ± 0.2	-13.9 ± 0.1	-58.65 ± 0.09	-58.62 ± 0.09
$M_{\text{p}}/M_\star$ [b]	$(3.4 \pm 0.8) \times 10^{-3}$	$(2.8 \pm 0.8) \times 10^{-3}$	$(4.1 \pm 0.8) \times 10^{-3}$	$(4.0 \pm 0.8) \times 10^{-3}$
$\mathbb{C}$ [c]	0.09 ± 0.02	0.10 ± 0.02	0.08 ± 0.02	0.08 ± 0.02
J_2	$(1.44 \pm 0.07) \times 10^{-4}$	$(1.66 \pm 0.08) \times 10^{-4}$	$(5.6 \pm 0.3) \times 10^{-5}$	$(6.1 \pm 0.3) \times 10^{-5}$
(Derived from the precession model)				
Precession period (yr)	$(5.7 \pm 0.4) \times 10^2$	$(4.3 \pm 0.3) \times 10^2$	$(1.6 \pm 0.2) \times 10^3$	$(1.5 \pm 0.2) \times 10^3$
L/S	$0.36^{+0.11}_{-0.09}$	0.25 ± 0.07	$0.65^{+0.24}_{-0.17}$	$0.54^{+0.19}_{-0.14}$

Note The quoted values and uncertainties are 50, 15.87, and 84.13 percentiles of the marginalized MCMC posteriors
[a] Value at BJD $= 2455633 = 2454833 + 800$
[b] Gaussian prior $M_{\text{p}}/M_\star = (4.2 \pm 0.8) \times 10^{-3}$ is imposed. The value is based on the average and standard deviation of the results by S14, Esteves et al. (2015), and Faigler and Mazeh (2015)
[c] Gaussian prior $\mathbb{C} = 0.0776 \pm 0.0200$ is imposed. The central value is from the result for $n = 3$ polytrope by Szabó et al. (2012) and the width is chosen to enclose that of the sun

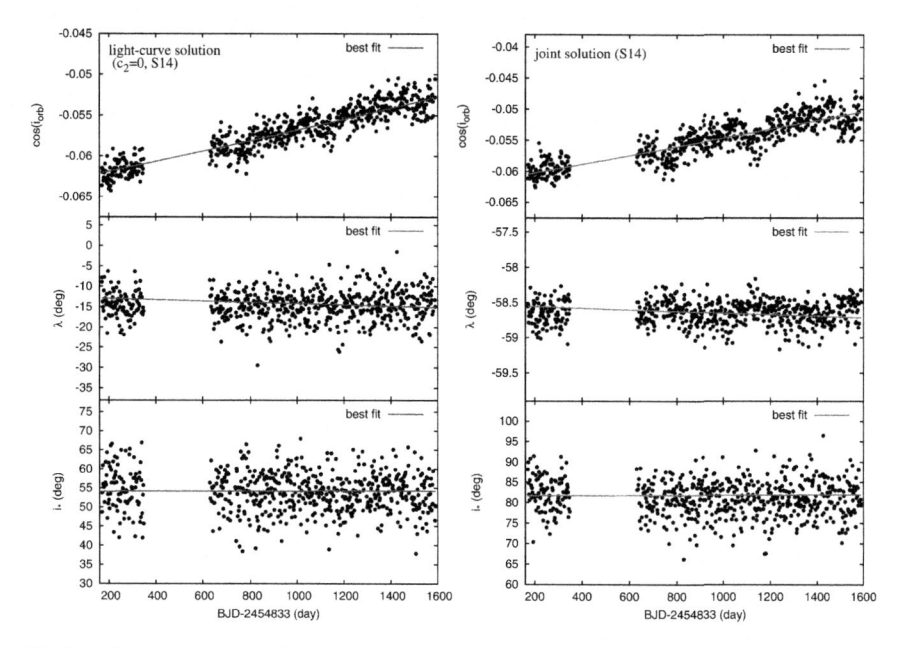

5.4 Spin–Orbit Precession in the Kepler-13A System

Fig. 5.4 Simultaneous fit to the observed $\cos i_{\rm orb}$, λ, and $i_\star$. (*Left*) light-curve solution with $c_2 = 0$. (*Right*) joint solution. Black points are from the light-curve fit (same as Fig. 5.3), and colored solid lines denote the best-fit precession models, which are *not* the linear fits

Basically, the precession model is compatible with the observations both for the light-curve solution and the joint solution. The value of J_2 and the corresponding precession period, however, are different by a factor of a few, in spite of the similar observed slopes in $\cos i_{\rm orb}$. While $J_2 = (1.66 \pm 0.08) \times 10^{-4}$ for the light-curve solution is consistent with the earlier estimate by Szabó et al. (2012), $J_2 = (2.1 \pm 0.6) \times 10^{-4}$ from observed TDVs and $J_2 = 1.7 \times 10^{-4}$ from the stellar model, the joint solution yields a smaller value, $J_2 = (6.1 \pm 0.3) \times 10^{-5}$.

The difference comes from the different three-dimensional architectures of the system described by the two solutions. Since all of $\cos i_{\rm orb}$, λ, and $i_\star$ are constrained from the gravity-darkened light curves, relative configuration of the stellar spin and orbital angular momenta are completely specified in three dimensions. This means that the *phase* of the precession during the *Kepler* mission, which corresponds to the left end in the right column of Fig. 5.5, is observationally constrained; from the top panel, we find that $\cos i_{\rm orb}$ is closer to the bottom of the sine curve for the light-curve solution (blue dashed line), while that for the joint solution (red solid line) resides in the phase of a rapid increase. For this reason, a larger precession rate (i.e., shorter precession period) is required for the light-curve solution to match the observed change in $\cos i_{\rm orb}$. According to Eqs. (5.5) and (5.6), the larger precession rate can be achieved by increasing either J_2 or L/S. However, the larger precession rate also induces faster variations in λ and $i_\star$, contradicting their almost constant observed values (middle and bottom panels in Fig. 5.4). The only way to mitigate this conflict is to make J_2 larger (i.e., increase the precession rate) while keeping L/S small, making it more difficult to move stellar spin axis. With Eq. (5.7), this explains why $M_p/M_\star$ is smaller and $\mathbb{C}$ is larger for the light-curve solution than for the joint solution in Table 5.2. Accordingly, the bottom panel of the right column in Fig. 5.5 exhibits the smaller precession amplitude for $i_\star$ in the former solution (blue dashed line) than the latter (red solid line).

The approximately three times difference in the precession period would be apparent even on the short time scale (left column in Fig. 5.5). As shown in the middle panel, as large as $\sim 10°$ change in λ is expected within the next ~ 10 yr for the light-curve solution, which is well detectable given the current precision of the sky-projected obliquity measurement with Doppler tomography (nominally down to a few degrees). On the other hand, λ for the joint solution is almost constant. From this point of view, the joint solution may slightly be favored even with the current data, because the nearly-constant values observed for λ and $i_\star$ are more natural for the joint solution than for the light-curve one, for the reasons discussed in the previous paragraph. This indication also manifests itself in the fact that the resulting $M_p/M_\star$ and $\mathbb{C}$ better agree with our prior knowledge in the joint solution.

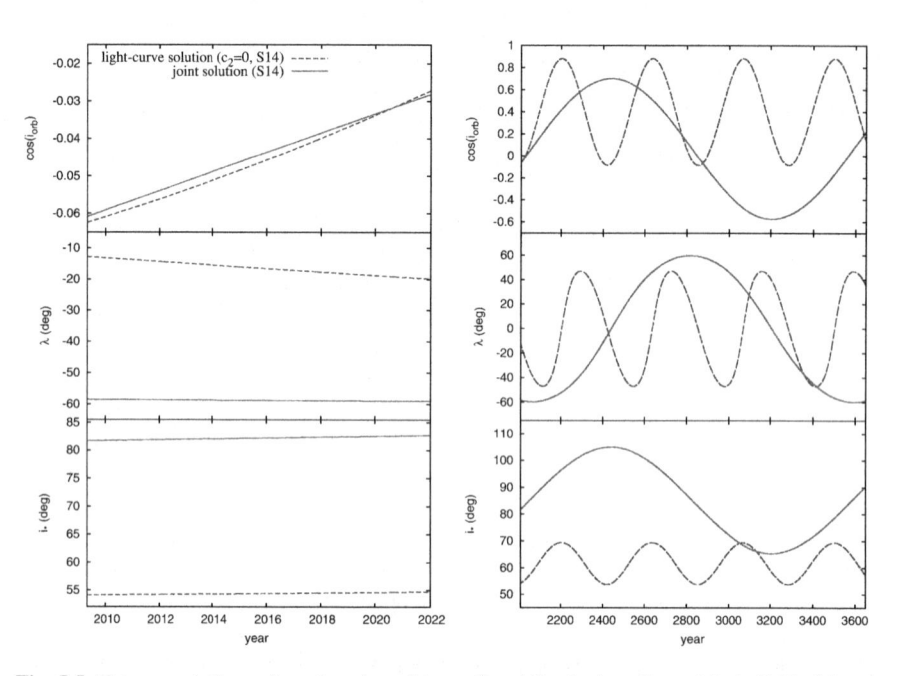

Fig. 5.5 Future evolutions of $\cos i_{\rm orb}$, λ, and $i_\star$ predicted for the best-fit models in Table 5.2 and Fig. 5.4. From top to bottom, the evolutions of $\cos i_{\rm orb}$, λ, and $i_\star$ are plotted for the solution from the light-curve alone (blue dashed line) and joint solution (red solid line) obtained with the S14 stellar parameters. The left panels show the short-term (~ 14 yr, until 2022) behavior, while the right ones are for the long-term (~ 1600 yr) variation

The more decisive conclusion will be obtained with the future follow-up observations of λ using Doppler tomography, as well as the transit duration observations to better constrain $\cos i_{\rm orb}$, and hence the precession rate. If our joint solution is correct, variations in λ will not be detected in near future. On the other hand, if the light-curve solution is actually correct and λ from the Doppler tomography is somehow systematically biased, λ should change; this *temporal variation* would be observable with the Doppler tomography even if it were biased. Or, it may even turn out that the precession scenario is wrong. In any case, tracking the future evolution of the system configuration can be used for an independent test of our solution, not to mention for better constraining stellar internal structure via J_2 and $\mathbb{C}$.

5.5 Anomaly in the Transit Light Curve of HAT-P-7

Armed with the methodology established using the distinct anomaly in Kepler-13A (Sect. 5.3), we discuss another, more subtle anomaly in this section. Here the methodology is further extended to include the information from asteroseismology as well as from the RM effect, and applied to an F-type star.

It has been pointed out in several studies, including the one in Chap. 4, that the transit light curve of HAT-P-7 exhibits a small anomaly of $\mathcal{O}(10^{-5})$. Morris et al. (2013), who reported this anomaly first, attributed it to the local spot-like gravity darkening induced by the gravity of the Jupiter-mass companion HAT-P-7b. They ruled out the gravity darkening of stellar rotational origin on the basis of the inspection that the anomaly is localized in a part of the transit. Later analyses with more data (e.g., Esteves et al. 2013, 2015; Van Eylen et al. 2013, and the one in Chap. 4), however, have shown that the anomaly is seemingly correlated over the whole transit duration, as in the top panel of Fig. 5.7. Moreover, the amplitude of the observed anomaly may be too large to be explained by the spot scenario. According to Jackson et al. (2012), the planet's gravity induces the surface temperature variation of "a few 0.1 K," which leads to the surface brightness variation of $\Delta F \sim$ several 100 ppm. If a planet crosses over a spot fainter by ΔF than the other part of the stellar disk, amplitude of the expected anomaly in the *relative flux* is about $\Delta F \times (R_{\rm p}/R_\star)^2 \sim \mathcal{O}({\rm ppm})$, which is order-of-magnitude smaller than the observed one. We therefore analyze this anomaly assuming that it is originated from the gravity darkening induced by stellar rotation, whose effect should not be localized but manifest during the whole transit duration.

Unlike the case of Kepler-13A, anomaly in the transit light curve is not clear on a quarter-by-quarter basis for HAT-P-7. In addition, no TTVs/TDVs have been detected for this planet. For these reasons, we deal with the light curve obtained by folding all the available SC, PDCSAP fluxes (Q0–17) processed as described in Sect. 5.2.2. We use the spectroscopic constraint $v \sin i_\star = 3.8 \pm 1.5 \,{\rm km\,s^{-1}}$ throughout this section. This value is based on Pál et al. (2008), though its error bar is enlarged to take into account other estimates for this quantity that give slightly different values (e.g., Winn et al. 2009a).

5.5.1 Robustness of the Observed Anomaly

If the observed anomaly is really due to gravity darkening, it should be persistent over all observation span, unlike the case of sporadic events including the spot crossing. It is important to confirm the property because Morris et al. (2013) only reported the bump before the mid-transit time. Thus, we divide the transits into four consecutive groups (Q0–4, 5–9, 10–13, 14–17), phase-fold and fit each of them with the model without gravity darkening separately, and examine the shapes of the residuals. Although fewer numbers of transits lead to noisier phase-folded light curves, ten-minute binned residuals in the left column of Fig. 5.6 exhibit a similar feature (brightening before mid-transit and dimming after it) in every span of data.

Besides, Van Eylen et al. (2013) reported seasonal variation in the transit depth depending on the quarter, which is reproduced in our analysis with Q0–17 data.[10] To confirm that the anomaly is not an artifact related to this seasonal variation, we also perform a similar analysis as above but this time grouping the transits that have similar depths. As shown in the right column of Fig. 5.6, we find that the same feature is apparent regardless of the season and the anomaly is not affected by the systematic depth variation. For this reason, along with its unconstrained origin, we do not try to make corrections for this systematic in the following analyses.

5.5.2 Results

As in Sect. 5.3, we consider both light-curve solution and joint solution that takes into account the constraints from other observations. First, the light-curve solution is obtained with c_2 fixed to be zero (Fig. 5.7, second and third columns in Table 5.3). In this case, we find two solutions with different signs of $\cos i_{\rm orb}$, which are indistinguishable in terms of the minimum χ^2.[11] The values quoted in Table 5.3 are the median, 15.87, and 84.13 percentiles of the MCMC posteriors sampled with emcee (Foreman-Mackey et al. 2013).[12] Our model reasonably reproduces the global feature of the anomaly (positive before the mid transit and negative after it), yielding $\Delta\chi^2 \simeq 166$ for ~ 420 degrees of freedom. We compute the Bayesian information

---

[10] We also reported a similar phenomenon in Kepler-13A; see Sect. 5.4.1 and Fig. 5.3.

[11] The existence of the two solutions in this case should be distinguished from the degeneracy intrinsic to the gravity-darkening method. For each of the two solution listed here, there additionally exists the model that yields exactly the same light curve, where $\cos i_{\rm orb}$ is replaced with $-\cos i_{\rm orb}$ and λ with $\pi - \lambda$. These intrinsically-degenerate solutions are not discussed here because they are in any case rejected in the joint solution, where λ is constrained by the prior. This is the same logic as used in the last paragraph of Sect. 5.3.1.

[12] We also applied the residual permutation method described in Winn et al. (2009b) for another estimate of the parameter uncertainties, and confirmed that they are not significantly affected by the correlated noise component.

5.5 Anomaly in the Transit Light Curve of HAT-P-7

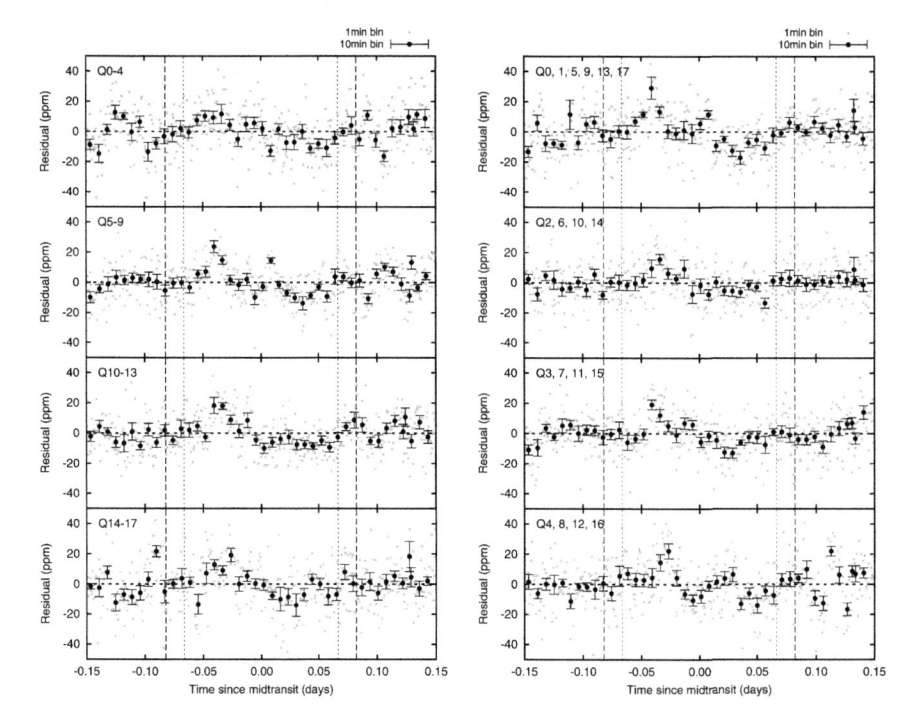

Fig. 5.6 Robustness of the detected anomaly. Residuals of the model fits (without gravity darkening) to the phase-folded transit light curves are plotted. Gray dots are residuals for the one-minute binned data, and black ones with error bars are the residuals averaged into ten-minutes bins. Vertical dashed and dotted lines correspond to the beginnings and ends of the ingress and egress. (Left column) Transits folded over different epochs. From top to bottom, light curves from Quarters 0–4, 5–9, 10–13, and 14–17 are folded. (Right column) Transits grouped by the CCD module used to observe the target. From top to bottom, light curves taken with the modules 17, 19, 9, and 7 are folded

criterion (BIC) for the best-fit models with and without gravity darkening, and find $\Delta \text{BIC} = 129$, which formally indicates that the gravity-darkened model is strongly favored.

Our solution points to a nearly pole-on configuration with $i_\star \simeq 0°$. This conclusion is consistent with the asteroseismic analyses in Chap. 4 (Benomar et al. 2014) and by Lund et al. (2014), but the nominal constraint on $i_\star$ from the gravity-darkened model is much tighter. On the other hand, λ is not constrained very well with the light curve asymmetry alone. The difficulty is inevitable in the pole-on configuration, where the brightness distribution on the stellar disk is almost axisymmetric even in the presence of gravity darkening. In such a case, ψ is always close to 90° regardless of λ.

One remaining issue regarding our solution is that the resulting rotation frequency may be too large. Given the age ($\simeq 2$ Gyr) and $B - V$ (= 0.495 ± 0.022; Lund et al. 2014) of the host star, the rotation frequency from the light-curve solution, $f_\text{rot} = 7.7 \pm 0.2$ mHz (equivalent to $P_\text{rot} \simeq 1.5$ days), is consistent with the gyrochronology relation by Meibom et al. (2009); see Sect. 6 of Lund et al. (2014). However, our value of f_rot is much larger than those from asteroseismology, $0.70^{+1.02}_{-0.43}$ mHz (68% credible interval obtained in Chap. 4) and < 0.8748 mHz (1σ upper limit by Lund et al. 2014). In fact, the prior used in these analyses, $|f_\mathrm{rot}| < 8$ mHz, does not fully cover the range we investigate here with the gravity-darkened light curve. Still, the discrepancy is only weakly reduced even with the new analysis adopting the prior range extended up to 17 mHz, which yields $f_\mathrm{rot} = 0.82^{+2.02}_{-0.50}$ mHz as the 68% credible interval (by courtesy of Othman Benomar; see also Benomar et al. 2014).

To examine if the gravity-darkened model is compatible with the seismic analysis, we then search for a joint solution including the constraints both from the RM measurement and asteroseismology. From the RM effect, we incorporate the constraint $\lambda = 172° \pm 32°$, which comes from the average and standard deviation of the analyses of the three different radial velocity data (Table 4.3 in Chap. 4). From asteroseismology, we adopt the above updated posterior for f_rot as the prior, and performed an MCMC sampling with emcee. To properly take into account the uncertainty from the limb-darkening profile, c_2 is also floated. The resulting credible intervals are summarized in the fourth and fifth columns in Table 5.3, and the model that maximizes the likelihood multiplied by the prior on f_rot is plotted with a dashed line in Fig. 5.7. The corresponding joint posterior distributions are also

Fig. 5.7 Fitting the gravity-darkened model to the phase-folded transit of HAT-P-7b. The meanings of the symbols are the same as those in Fig. 5.1, but this time the joint solution incorporates the constraints on λ from the RM measurement and on f_rot from asteroseismology. The light-curve solution and joint solution are almost indistinguishable in this case, as expected from the similar values of χ^2

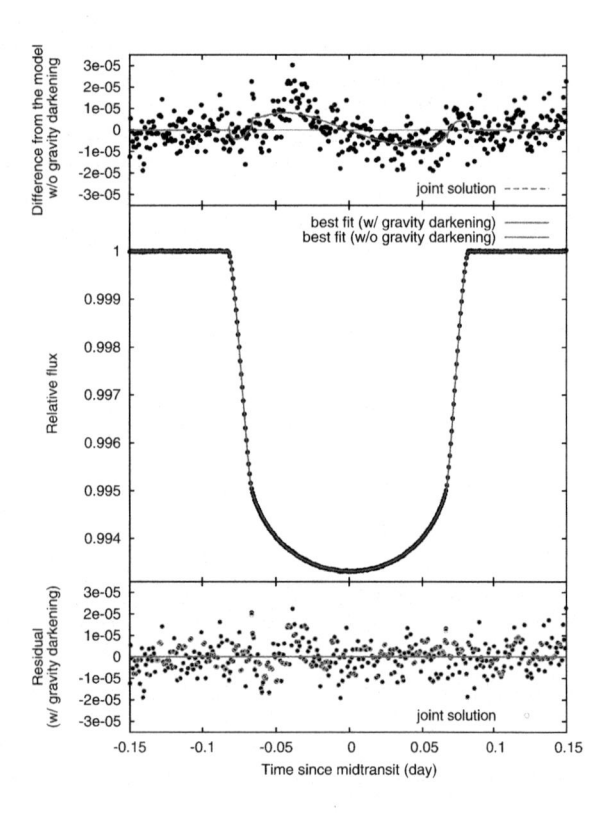

shown in Figs. C.7 and C.8. We again find two equally good solutions, both of which indicate nearly pole-on configurations with slightly prograde and retrograde orbits, $\psi = 101° \pm 2°$ and $\psi = 87° \pm 2°$. Although the resulting f_{rot} still prefers a higher rotation rate than that from asteroseismology, their difference is now mitigated to the 2σ level: we construct the probability distribution for Δf_{rot}, f_{rot} from out joint analysis minus f_{rot} from asteroseismology, using their posteriors and find its 2σ credible region as $\Delta f_{rot} = 4.9^{+4.0}_{-5.0}\,\mu\text{Hz}$. We argue that the level of discrepancy is acceptable, considering that the rotational mode splitting is not clearly detected in the power spectrum for HAT-P-7.

Finally, it is also worth considering the case where $\beta \neq 0.25$, given the unconstrained nature of the gravity darkening in F dwarfs. Smaller values of $\beta \sim 0.08$ are usually expected for solar-like stars with convective envelopes (e.g., Lucy 1967; Claret 1998), while Lara and Rieutord (2011, 2012) argue that β is close to 0.25 in the limit of slow rotation under several assumptions. We repeat the above joint analysis floating β with the prior uniform between 0 and 0.3, and obtain $\beta = 0.26^{+0.03}_{-0.05}$ for both solutions in Table 5.3. On the one hand, the fact may support the claims by Lara and Rieutord (2011, 2012); on the other hand, it may simply indicate some incompleteness in our gravity-darkening model, as also suggested by the tension in f_{rot} and the still correlated residuals before the mid transit (bottom panel of Fig. 5.7). Indeed, if $\beta = 0.08$ is adopted, we find that even higher rotation rate ($>10\,\text{mHz}$) is favored, making the discrepancy with asteroseismology more serious. Although the validity of β we obtain is beyond the scope of this chapter, we note that our conclusion for a pole-on orbit is robust against the adopted value of β; in both analyses where β is fitted and β is fixed to be 0.08, the constraints on ψ differ less than 1σ from the results in Table 5.3.

5.6 Summary

5.6.1 Kepler-13A

First, we analyze the gravity-darkened transit light curve of Kepler-13A adopting the same model and stellar parameters as in the previous study by B11. We reproduce the spin–orbit angles obtained by B11 with more data (called "light-curve solution" in this chapter) and also find that the choice of the stellar mass, stellar effective temperature, $v \sin i_\star$, or contaminated flux affects λ or $i_\star$ by less than about 10°. If we fit $c_2 = u_1 - u_2$ as well as $c_1 = u_1 + u_2$ in the quadratic limb-darkening law, on the other hand, a broader range of the spin–orbit angle is allowed. In fact, this additional degree of freedom may explain the discrepancy between the solution by B11 and the Doppler tomography result by Johnson et al. (2014). Our new "joint solution" includes $i_\star = 81° \pm 5°$, $\lambda = -59° \pm 2°$, $\psi = 60° \pm 2°$, and $P_{rot} = 24 \pm 2\,\text{hr}$. Although the joint solution is compatible with all of the observations made so far, introducing additional free parameter c_2 is not statistically justified, nor is it clear if the best-fit value for c_2 is physically plausible.

To examine the above issues from a dynamical point of view, we also analyze the spin–orbit precession in this system. By analyzing the light curves from each

Table 5.3 Results for the transit of HAT-P-7b

	Light-curve solution ($c_2 = 0$)		Joint solution (c_2 fitted)	
	Solution 1	Solution 2	Solution 1	Solution 2
(Constraints)				
$v \sin i_\star$ (km s^{-1})	3.8 ± 1.5	3.8 ± 1.5	3.8 ± 1.5	3.8 ± 1.5
λ (deg)	...	...	172 ± 32	172 ± 32
(Fitted parameters)				
$M_\star$ ($M_\odot$)	1.59 (fixed)	1.59 (fixed)	1.59 (fixed)	1.59 (fixed)
$T_{\star,\text{pole}}$ (K)	6310 (fixed)	6310 (fixed)	6310 (fixed)	6310 (fixed)
$\rho_\star$ (g cm^{-3})	0.2789 ± 0.0006	0.2789 ± 0.0006	0.2790 ± 0.0005	0.2784 ± 0.0005
c_1	0.498 ± 0.003	0.498 ± 0.003	$0.507^{+0.008}_{-0.016}$	$0.508^{+0.007}_{-0.015}$
c_2	0 (fixed)	0 (fixed)	$0.07^{+0.06}_{-0.12}$	$0.08^{+0.05}_{-0.11}$
t_c (10^{-5} day)[a]	-1.5 ± 0.4	-1.5 ± 0.4	-1.6 ± 0.4	-1.1 ± 0.4
P (day)	2.204735471 (fixed)			
$\cos i_{\text{orb}}$	-0.1195 ± 0.0004	0.1195 ± 0.0004	-0.1194 ± 0.0003	0.1198 ± 0.0003
$R_p/R_\star$	$0.07757^{+0.00005}_{-0.00009}$	$0.07757^{+0.00005}_{-0.00009}$	0.07759 ± 0.00003	$0.07749^{+0.00003}_{-0.00004}$
F_0	0.9999998 ± 0.0000005			
f_{rot} (mHz)	7.7 ± 0.2	7.7 ± 0.2	$6.1^{+2.6}_{-1.7}$[b]	$5.6^{+2.4}_{-1.7}$[b]
$i_\star$ (deg)[c]	$3.3^{+1.2}_{-1.0}$	$3.3^{+1.3}_{-1.0}$	$5.3^{+3.3}_{-2.0}$	$5.3^{+3.7}_{-2.1}$
λ (deg)	133^{+19}_{-88}	49^{+92}_{-21}	142^{+12}_{-16}	136^{+16}_{-22}
β	0.25 (fixed)	0.25 (fixed)	0.25 (fixed)	0.25 (fixed)
(Derived parameters)				
P_{rot} (day)	1.51 ± 0.03	1.51 ± 0.03	$1.9^{+0.7}_{-0.6}$	$2.1^{+0.9}_{-0.6}$
ψ (deg)	99^{+2}_{-4}	81^{+4}_{-2}	101 ± 2	87 ± 2
Impact parameter	0.496 ± 0.001	0.496 ± 0.001	0.496 ± 0.001	0.497 ± 0.001
Oblateness	0.0149 ± 0.0006	0.0149 ± 0.0007	$0.009^{+0.010}_{-0.005}$	$0.008^{+0.008}_{-0.004}$
$\chi^2_{\min}/\text{dof}$	453/424	455/424	450/424	451/424

Note The quoted values and uncertainties are 50, 15.87, and 84.13 percentiles of the marginalized MCMC posteriors. For the light-curve solution, $\chi^2_{\min}$ is the value of χ^2 computed from Eq. (5.4) for the maximum likelihood model. Eq. (5.4) is also used for the joint solution, but $\chi^2_{\min}$ in this case is computed for the model that maximizes the likelihood function multiplied by the prior on f_{rot}

[a] Measured from the transit epoch $t_0(\text{BJD} - 2454833) = 120.358522 \pm 0.000005$ obtained with the transit model without gravity darkening

[b] Posterior from the seismic analysis is used as the prior

[c] We impose the prior uniform in $\cos i_\star$, rather than in $i_\star$, which corresponds to the isotropic distribution for the spin direction

quarter separately, we confirm that the variation in $|\cos i_\mathrm{orb}|$ causes the transit duration variations first reported by Szabó et al. (2012), with more elaborate model taking into account the gravity darkening. This variation is consistent with the precession of the stellar spin and orbital angular momenta around the total angular momentum of the system, induced by the oblateness of the rapidly rotating host star. We thus fit each transit with the gravity-darkened model to determine $\cos i_\mathrm{orb}$, λ, and $i_\star$ as a function of time, and then fit them with the precession model to constrain the stellar quadrupole moment J_2. For the light-curve solution and the joint solution, we respectively find $J_2 = (1.66 \pm 0.08) \times 10^{-4}$ and $J_2 = (6.1 \pm 0.3) \times 10^{-5}$, which are different by a factor of a few. Our results predict detectable variations in λ on 10-yr timescale for the light-curve solution, while it should be almost constant for the joint solution. The difference suggests that the future follow-up observations can be used to confirm or refute the joint solution we proposed, as well as to improve the constraint on J_2.

5.6.2 HAT-P-7

Although the anomaly in the transit light curve is much more subtle compared to Kepler-13Ab, we confirm that the asymmetric residual (not only the bump reported by Morris et al. (2013) but also the dip) exists continuously in the transits of HAT-P-7b. Thus, we perform the analysis assuming that the gravity-darkening is a viable explanation for the anomaly. Gravity-darkened transit model favors a nearly pole-on orbit ($\psi = 101° \pm 2°$ or $\psi = 87° \pm 2°$) and the gravity-darkening exponent β close to 0.25, consistently with the asteroseismic inference in Chap. 4. The constraint on ψ is insensitive to the choice of the limb-darkening parameters or the gravity-darkening exponent.

On the other hand, the stellar rotation rate from the gravity-darkening analysis is about 2σ higher than the value from asteroseismology. In addition, the value of $\beta \simeq 0.25$ we obtained may be too large for a star with a convective envelope, although the theory of gravity darkening may not be full-fledged for that case. These facts may suggest some incompleteness in the current modeling or other origins for the anomaly, and should be addressed in future studies.

5.7 Conclusion

Our present analysis reproduces the results by B11 with more data and thus strengthens the reliability of the gravity-darkening method for the spin–orbit angle determination. In contrast, we also find that the spin–orbit angle obtained from the gravity-darkened transit light curve strongly depends on the assumed limb-darkening profile. Depending on its choice, the resulting spin–orbit angle can vary by several tens of degrees. Thus, the reliable modeling of the limb-darkening effect is crucial for this method.

Nevertheless, if λ is constrained from other observations, $i_\star$ is well determined along with the limb-darkening parameters. Hence the gravity-darkening method still provides valuable information on the true stellar obliquity ψ, which is complementary to λ from the RM effect or Doppler tomography. Indeed, such an example is already seen in an eclipsing binary system DI Her (Philippov and Rafikov 2013). In addition, synergy with asteroseismology is also promising because it constraints $f_{\rm rot}$ and $i_\star$, which are both essential in the modeling of gravity darkening. The joint analyses of these kinds may in turn help us to better understand the mechanisms of gravity darkening itself, since they enable the measurements of β for stars not in close binary systems and hence free from the strong tidal distortion.

If combined with continuous, high-precision photometry as achievable with space-borne instruments, the gravity-darkening method also provides a way to monitor the angular momentum evolution in the system. Modeling of the spin–orbit precession allows us to access the internal structure of the rotating star through its quadrupole moment or moment of inertia. It is also possible to precisely determine the three-dimensional configuration of the system from a dynamical point of view (cf. Philippov and Rafikov 2013; Barnes et al. 2013). Such information will be valuable in simulating the dynamical histories of individual systems to decipher the origin of the spin–orbit misalignment.

References

J.P. Ahlers, S.A. Seubert, J.W. Barnes, ApJ **786**, 131 (2014)
S. Albrecht, J.N. Winn, J.A. Johnson et al., ApJ **757**, 18 (A12) (2012)
J.W. Barnes, E. Linscott, A. Shporer, ApJs **197**, 10 (B11) (2011)
J.W. Barnes, ApJ **705**, 683 (2009)
J.W. Barnes, J.C. van Eyken, B.K. Jackson, D.R. Ciardi, J.J. Fortney, ApJ **774**, 53 (2013)
M.R. Bate, G. Lodato, J.E. Pringle, MNRAS **401**, 1505 (2010)
O. Benomar, K. Masuda, H. Shibahashi, Y. Suto, PASJ **66**, 94 (2014)
W.J. Borucki, D.G. Koch, G. Basri et al., ApJ **736**, 19 (2011)
G. Boué, J. Laskar, Icarus **201**, 750 (2009)
A. Claret, A&AS **131**, 395 (1998)
L.J. Esteves, E.J.W. De Mooij, R. Jayawardhana, ApJ **772**, 51 (2013)
L.J. Esteves, E.J.W. De Mooij, R. Jayawardhana, ApJ **804**, 150 (2015)
S. Faigler, T. Mazeh, ApJ **800**, 73 (2015)
D.B. Fielding, C.F. McKee, A. Socrates, A.J. Cunningham, R.I. Klein, MNRAS **450**, 3306 (2015)
D. Foreman-Mackey, D.W. Hogg, D. Lang, J. Goodman, PASP **125**, 306 (2013)
B.K. Jackson, N.K. Lewis, J.W. Barnes et al., ApJ **751**, 112 (2012)
M.C. Johnson, W.D. Cochran, S. Albrecht et al., ApJ **790**, 30 (2014)
S. Kamiaka, K. Masuda, Y. Xue et al., PASJ **67**, 94 (2015)
D. Lai, F. Foucart, D.N.C. Lin, MNRAS **412**, 2790 (2011)
F.E. Lara, M. Rieutord, A&A **533**, A43 (2011)
F.E. Lara, M. Rieutord, A&A **547**, A32 (2012)
G. Li, S. Naoz, F. Valsecchi, J.A. Johnson, F.A. Rasio, ApJ **794**, 131 (2014)
L.B. Lucy, ZAp **65**, 89 (1967)
M.N. Lund, M. Lundkvist, V. Silva Aguirre et al., A&A **570**, A54 (2014)
K. Mandel, E. Agol, ApJ **580**, L171 (2002)

References

C.B. Markwardt, *Astronomical Data Analysis Software and Systems XVIII*, ed. by D.A. Bohlender, D. Durand, P. Dowler. Astronomical Society of the Pacific Conference Series, vol. 411 (2009), p. 251

S. Meibom, R.D. Mathieu, K.G. Stassun, ApJ **695**, 679 (2009)

J.D. Monnier, M. Zhao, E. Pedretti et al., Science **317**, 342 (2007)

B.M. Morris, A.M. Mandell, D. Deming, ApJ **764**, L22 (2013)

N. Narita, B. Sato, T. Hirano, M. Tamura, PASJ **61**, L35 (N09) (2009)

A. Pál, G.Á. Bakos, G. Torres et al., ApJ **680**, 1450 (P08) (2008)

A.A. Philippov, R.R. Rafikov, ApJ **768**, 112 (2013)

A. Shporer, J.G. O'Rourke, H.A. Knutson et al., ApJ **788**, 92 (S14) (2014)

D.K. Sing, A&A **510**, A21 (2010)

J. Southworth, MNRAS **386**, 1644 (2008)

N.I. Storch, K.R. Anderson, D. Lai, Science **345**, 1317 (2014)

G.M. Szabó, R. Szabó, J.M. Benkő et al., ApJ **736**, L4 (2011)

G.M. Szabó, A. Pál, A. Derekas et al., MNRAS **421**, L122 (2012)

G.M. Szabó, A. Simon, L.L. Kiss, MNRAS **437**, 1045 (2014)

V. Van Eylen, M. Lindholm Nielsen, B. Hinrup, B. Tingley, H. Kjeldsen, ApJ **774**, L19 (2013)

H. von Zeipel, MNRAS **84**, 665 (1924)

J.N. Winn, J.A. Johnson, S. Albrecht et al., ApJ **703**, L99 (W09) (2009a)

J.N. Winn, M.J. Holman, G.W. Henry et al., ApJ **693**, 794 (2009b)

J.N. Winn, D. Fabrycky, S. Albrecht, J.A. Johnson, ApJ **718**, L145 (2010)

Y. Xue, Y. Suto, A. Taruya et al., ApJ **784**, 66 (2014)

G. Zhou, C.X. Huang, ApJ **776**, L35 (2013)

Chapter 6
Probing the Architecture of Hierarchical Multi-Body Systems: Photometric Characterization of the Triply-Eclipsing Triple-Star System KIC 6543674

Abstract It is mainly due to the unknown initial distribution of stellar obliquities that the spin–orbit misalignment cannot be immediately interpreted as a signature of past dynamical interactions. In contrast, the misalignment between planetary orbits in the same system will serve as more direct evidence for the past dynamical event and may also complement the interpretation of obliquity measurements. Indeed, the scenarios for the high-eccentricity migration discussed in Sect. 3.1 all involve the excitation of the mutual orbital inclination. Moreover, these processes themselves are more generic than the hot Jupiter formation and their signature may also be observed in systems without hot Jupiters. Such systems, if identified and characterized in detail, will serve as direct evidence for the dynamical scenario and/or useful test beds for studying how it works in real systems. The high-eccentricity scenario often involves a hierarchical configuration (see Sect. 3.1), where the orbit of an outer planet/star is much wider than that of the inner planet. For this reason, characterization of hierarchical systems will be a key for the purpose described above. As the first step of such an effort, this chapter presents the characterization of a hierarchical triple-star system based on the *Kepler* photometric data. We determine the three-dimensional orbits and physical dimensions of all three stars in the system by a joint modeling of the eclipse light curves and mutual gravitational interaction: the technique also applicable to a hierarchical planetary system with a massive outer planet. We also discuss the implication for the very close inner orbit of this system, whose origin may be similar to those of hot Jupiters.

Keywords Close binary · Eclipse timing variations
Kozai cycles with tidal friction · KIC 6543674

6.1 Introduction

Among over 2000 eclipsing binaries discovered in the *Kepler* mission (Prša et al. 2011; Slawson et al. 2011), more than 200 are suggested to host tertiary (third body) companions through their eclipse timing variations (ETVs; Conroy et al. 2014).

Many of them are hierarchical triples consisting of a short-period binary and an outer third body on a wide orbit. The hierarchy is often attributed to the perturbation from the third body, as in the KCTF (Kozai cycles with tidal friction) scenario (Kozai 1962; Kiseleva et al. 1998; Eggleton and Kiseleva-Eggleton 2001) described in Sect. 3.1.1. Indeed, recent ETV analyses (Rappaport et al. 2013; Borkovits et al. 2015) have revealed many hierarchical triples with misaligned tertiary orbits, whose mutual inclination distribution exhibits a suggestive peak around ∼40° as predicted by the KCTF scenario (Fabrycky and Tremaine 2007).

On the other hand, at least 10 or more hierarchical triples seem to have well-aligned orbits, as suggested by eclipses due to tertiary companions (Carter et al. 2011; Orosz 2015, Fig. 7). Three-dimensional geometry and absolute dimensions of those systems are also of interest because their hierarchy may argue for mechanisms of orbital shrinkage that do not require high mutual inclinations between the inner and outer binary planes (e.g., Petrovich 2015).

In this chapter, we focus on a tertiary event observed only once in the KIC 6543674 system, which involves three tertiary eclipses around a single inferior conjunction

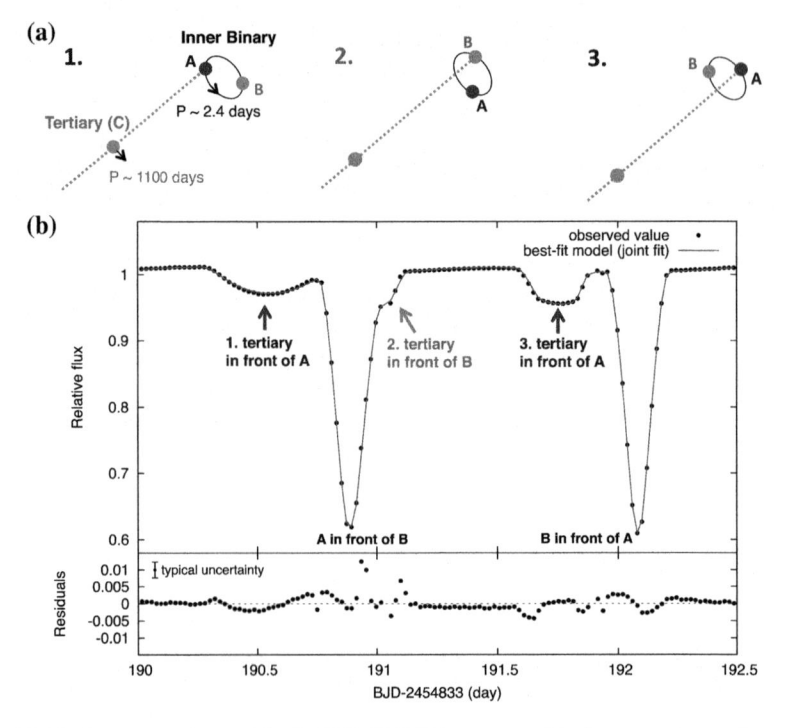

Fig. 6.1 Tertiary event observed in the KIC 6543674 system and its interpretation. **a** Schematic illustration of the system configuration during the event. **b** Fit to the *Kepler* light curve around the tertiary eclipses (see Sect. 6.3). (*Top*) Black circles are the observed fluxes and red solid line denotes our best-fit model. (*Bottom*) Residuals of our fit. Typical uncertainty estimated from our analysis ($\simeq \sigma_{\text{LC,tertiary}}$) is shown at the upper left

of the third body (Fig. 6.1). Although this event has already been reported (Slawson et al. 2011; Thackeray-Lacko et al. 2013; Conroy et al. 2014), it has not yet been clarified whether it is indeed explained by the tertiary eclipse, nor what information is obtained from its detailed modeling. Below we will show that the tertiary event plays two crucial roles in determining the system configuration. First, it constrains the mutual inclination between the inner and outer binary orbits very precisely, in a similar way to the "planet–planet eclipse" known in the *Kepler* multi-transiting planetary system(s) (Hirano et al. 2012; Masuda et al. 2013; Masuda 2014). Secondly, and less trivially, it fixes the mass ratio of the inner binary and velocity of the third body even without spectroscopy.

The present chapter reports precise geometry and absolute dimensions of the KIC 6543674 system. We combine the above information from the tertiary event with the complementary constraints from ETVs and eclipses of the inner binary. To obtain a consistent solution, we fit the three components simultaneously using a Markov Chain Monte Carlo (MCMC) method. Section 6.2 presents individual analyses of the ETVs and eclipse curves of the inner binary. We then model the two components jointly with the tertiary eclipses in Sect. 6.3 to determine the parameters of the whole system. Section 6.4 discusses the implication of the resulting system architecture and the prospects for the follow-up observations to better understand this valuable system.

6.2 Constraints from ETVs and Phase Curve of the Inner Binary

The KIC 6543674 system consists of the inner eclipsing binary with the orbital period of $P_{in} \simeq 2.39$ days and outer eccentric binary with $P_{out} \simeq 1100$ days; here the "outer" binary refers to the "binary" system consisting of the third body and the center of mass of the inner binary. In this section, we present individual MCMC analyses of the phase curve and ETVs of the inner binary, which allow us to constrain the orbital geometries of the inner and outer binaries, respectively. Since P_{in}/P_{out} is small, both inner and outer binary orbits are approximately Keplerian. We adopt the approximation throughout the chapter and define all the orbital elements in Jacobi coordinates (with subscripts "in" and "out"), which are in this case constant over time.

6.2.1 ETV Analysis

The inner binary exhibit ETVs, which were used to infer the existence of the third body in this system (Conroy et al. 2014). They are caused by the finite light-travel time (Rømer delay) and the variation in the line-of-sight distance due to the outer binary motion. Under our assumption, the ith eclipse time of the inner binary t_i can be modeled as (Rappaport et al. 2013)[1]

$$t_i^{\text{model}} = t_{0,\text{in}} + P_{\text{in}} i + A_{\text{ETV}} \left\{ \sqrt{1 - e_{\text{out}}^2} \sin E_{\text{out}}(t_i) \cos \omega_{\text{out}} + [\cos E_{\text{out}}(t_i) - e_{\text{out}}] \sin \omega_{\text{out}} \right\}. \tag{6.1}$$

Here, $t_{0,\text{in}}$ is the eclipse epoch (time of inferior conjunction) of the inner binary, and e_{out}, ω_{out}, and E_{out} are the eccentricity, argument of pericenter, and eccentric anomaly of the third body. The amplitude of ETVs, A_{ETV}, is given by the projected semi-major axis of the outer binary $a_{\text{out}} \sin i_{\text{out}}$ divided by the speed of light c:

$$A_{\text{ETV}} = \frac{(GM_A)^{1/3}}{c(2\pi)^{2/3}} \frac{(M_C/M_A) \sin i_{\text{out}}}{(1 + M_B/M_A + M_C/M_A)^{2/3}} P_{\text{out}}^{2/3}, \tag{6.2}$$

where M denotes the stellar mass, with the subscripts A, B, and C specifying the primary, secondary, and tertiary stars, respectively. In such a hierarchical system as KIC 6543674, dynamical effects that change P_{in} are sufficiently smaller than the above effect and so are neglected (Rappaport et al. 2013).

We use Eq. (6.1) to model the primary eclipse times t_i^{obs} in table 1 of Conroy et al. (2014) obtained by fitting the light curve over the entire phase (flagged as "entire"). The observed ETVs also exhibit short-term modulations (see Fig. 6.2a), which can be explained by star spots if the stellar rotation is nearly (but not exactly) synchronized with the inner binary motion (see, e.g., Fig. 3 of Orosz 2015). Instead of modeling them, we include an additional scatter σ_{ETV} to the formal eclipse-time error σ_i in quadrature to define the following likelihood for the ETV fit:

$$\mathcal{L}_{\text{ETV}} = \prod_i \frac{1}{\sqrt{2\pi(\sigma_i^2 + \sigma_{\text{ETV}}^2)}} \exp\left[\frac{(t_i^{\text{obs}} - t_i^{\text{model}})^2}{2(\sigma_i^2 + \sigma_{\text{ETV}}^2)}\right]. \tag{6.3}$$

By optimizing σ_{ETV} along with the other physical model parameters and marginalizing over it, we can obtain more realistic constraints taking into account the additional variation due to star spots. The likelihood in Eq. (6.3) is used to perform an MCMC sampling (emcee by Foreman-Mackey et al. 2013) of the posteriors of the parameters in the second column of Table 6.1. The best-fit model is compared with the observed values in Fig. 6.2a.

---

[1] The sign is opposite to their Eq. (6) because we take $+z$-axis in the observer's direction.

6.2 Constraints from ETVs and Phase Curve of the Inner Binary

Table 6.1 System parameters from the *Kepler* light curves

Parameter	ETVs	phase curve	ETVs + phase + tertiary	ETVs + phase + tertiary (with the prior on M_A)
(Inner binary)				
$t_{0,\mathrm{in}}$ (BJD − 2454833)	132.3070 ± 0.0002	...	132.3071 ± 0.0001	132.30704 ± 0.00009
$t_{0,\mathrm{in}}^{\mathrm{phase}}$ (BJD − 2454833)	...	132.30372 ± 0.00004	132.30372 ± 0.00003	$132.30372^{+0.00002}_{-0.00003}$
P_{in} (day)	2.3910305 ± 0.0000003	2.3910305 (fixed)	2.3910305 ± 0.0000003	2.3910305 ± 0.0000002
a_{in}/R_A	...	5.49 ± 0.02	$5.494^{+0.007}_{-0.006}$	$5.494^{+0.006}_{-0.007}$
$\cos i_{\mathrm{in}}$	...	0.021 ± 0.002	0.022 ± 0.002	0.022 ± 0.002
$e_{\mathrm{in}} \cos \omega_{\mathrm{in}}$	...	$(0.2 \pm 3.3) \times 10^{-5}$	0 (fixed)	0 (fixed)
$e_{\mathrm{in}} \sin \omega_{\mathrm{in}}$	...	$-0.0005^{+0.0021}_{-0.0020}$	0 (fixed)	0 (fixed)
R_B/R_A	...	0.781 ± 0.004	0.781 ± 0.002	0.781 ± 0.002
M_B/M_A	...	...	0.93 ± 0.02	0.93 ± 0.02
C_{phase}	...	1.00259 ± 0.00002	1.00259 ± 0.00002	1.00259 ± 0.00002
T_B/T_A	...	1.012 ± 0.002	1.0107 ± 0.0004	1.0107 ± 0.0004
u_A	...	0.45 ± 0.04	0.434 ± 0.009	0.434 ± 0.009
u_B	...	0.46 ± 0.03	0.47 ± 0.02	0.46 ± 0.02
A_0	...	0.041 ± 0.007	0.037 ± 0.006	0.037 ± 0.006
A_{1c}	...	0.00034 ± 0.00005	0.00035 ± 0.00005	0.00035 ± 0.00005
A_{1s}	...	0.00096 ± 0.00004	0.00096 ± 0.00004	0.00096 ± 0.00004
A_{2c}	...	−0.00720 ± 0.00007	−0.00716 ± 0.00006	−0.00716 ± 0.00006
$R_A\ (R_\odot)$	...	...	$2.1^{+3.2}_{-0.8}{}^{\dagger}$	$1.8 \pm 0.1^{\dagger}$
$R_B\ (R_\odot)$	...	...	$1.6^{+2.5}_{-0.7}{}^{\dagger}$	$1.4 \pm 0.1^{\dagger}$

(continued)

Table 6.1 (continued)

Parameter	ETVs	phase curve	ETVs + phase + tertiary	ETVs + phase + tertiary (with the prior on M_A)
M_A ($M_\odot$)	...	...	$1.8^{+27.5}_{-1.4}$†	1.2 ± 0.3
M_B ($M_\odot$)	...	...	$1.7^{+25.5}_{-1.3}$†	$1.1^{+0.3}_{-0.2}$†
(Third body)				
$t_{0,\text{out}}$ (BJD − 2454833)	199 ± 10	...	191.246 ± 0.003	191.246 ± 0.003
P_{out} (day)	1086^{+8}_{-7}	...	1090 ± 6	1090 ± 5
$e_{\text{out}} \cos \omega_{\text{out}}$	0.13 ± 0.05	...	0.16 ± 0.03	0.16 ± 0.03
$e_{\text{out}} \sin \omega_{\text{out}}$	0.58 ± 0.03	...	0.58 ± 0.02	0.572 ± 0.008
a_{out}/R_A	...	...	345^{+15}_{-13}	348 ± 2†
$\cos i_{\text{out}}$	...	...	0.0030 ± 0.0003	$0.0029^{+0.0001}_{-0.0002}$
$\Delta\Omega$ (deg)	...	...	3.2 ± 0.6	3.1 ± 0.6
A_{ETV} (s)	264 ± 6	...	266 ± 5	265 ± 5†
C_{tertiary}	...	...	1.0070 ± 0.0003	1.0070 ± 0.0003
γ_{tertiary} (day^{-1})	...	...	0.00004 ± 0.00021	$0.00005^{+0.00021}_{-0.00022}$
R_C/R_A	...	...	0.277 ± 0.003	0.277 ± 0.003
M_C/M_A	...	...	$0.4^{+0.3}_{-0.2}$†	$0.43^{+0.04}_{-0.03}$
T_C/T_A	...	...	$0.84^{+0.03}_{-0.04}$†	$0.84^{+0.03}_{-0.04}$†
R_C ($R_\odot$)	...	...	$0.6^{+0.9}_{-0.2}$†	0.50 ± 0.04†
M_C ($M_\odot$)	...	...	$0.7^{+3.3}_{-0.4}$†	$0.50^{+0.07}_{-0.08}$†
mutual inclination (deg)	...	...	3.3 ± 0.6†	$3.3^{+0.5}_{-0.6}$†
(Jitters)				
σ_{ETV} (s)	56 ± 3	...	56 ± 3	56 ± 3

(continued)

6.2 Constraints from ETVs and Phase Curve of the Inner Binary

Table 6.1 (continued)

Parameter	ETVs	phase curve	ETVs + phase + tertiary	ETVs + phase + tertiary (with the prior on M_A)
$\sigma_{LC,phase}$	...	0.00048 ± 0.00001	0.00049 ± 0.00001	0.00049 ± 0.00001
$\sigma_{LC,tertiary}$	...	...	0.0023 ± 0.0002	$0.0023^{+0.0002}_{-0.0001}$

Note The quoted values and uncertainties are the median and 68.3% credible interval of the marginalized posteriors. Values marked with daggers are derived from the posteriors of other fitted parameters

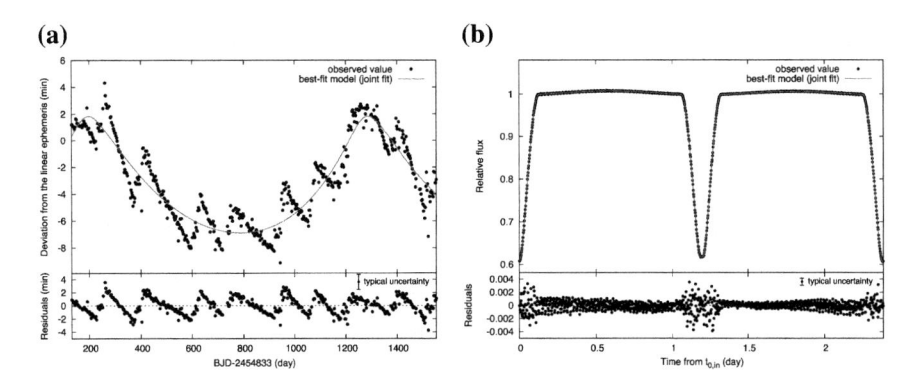

Fig. 6.2 **a** Fit to the eclipse times. (*Top*) Black circles are the observed eclipse times and red solid line denotes our best-fit model. Only the deviations from the linear ephemeris, i.e., *variations* in the eclipse times, are shown for clarity. (*Bottom*) Residuals of our fit. Typical (jitter-included) uncertainty is shown at the upper right. **b** Fit to the folded phase curve. (*Top*) Black circles are the observed fluxes and red solid line denotes our best-fit model. (*Bottom*) Same as panel (**a**)

6.2.2 Phase-Curve Analysis

The linear ephemeris of the inner binary ($t_{0,in}$ and P_{in}) obtained in Sect. 6.2.1 is used to phase-fold the light curve taken from the *Kepler* eclipsing binary catalog,[2] whose instrumental trend has been removed ("flattened") using polynomials (Conroy et al. 2014). Since A_{ETV} is shorter than the data cadence (29.4 min), we do not correct for ETVs here and in the following light-curve fitting (Sect. 6.3). The folded fluxes are averaged into three minute bins, and the flux value and error in each bin are estimated as the median and 1.4826 times median absolute deviation divided by the square root of the number of points in the bin.

---

[2] http://keplerebs.villanova.edu.

We model the flux over the entire phase as

$$f_{\text{phase}}(t) = \frac{C_{\text{phase}}}{1 + F_B/F_A + A_0} \left[f_A(t) + \frac{F_B}{F_A} f_B(t) + A_0 + A_{1c} \cos\phi + A_{1s} \sin\phi + A_{2c} \cos 2\phi \right]. \tag{6.4}$$

Here, $f_{A,B}(t)$ is the normalized stellar flux computed with the analytic eclipse model by Mandel and Agol (2002) for the linear limb darkening law. They are determined from the orbital ephemeris, scaled semi-major axis a_{in}/R_A, cosine of the orbital inclination $\cos i_{\text{in}}$, radius ratio R_B/R_A, and linear limb-darkening coefficients u_A and u_B. The flux ratio, F_B/F_A, is computed by $(R_B/R_A)^2 (T_B/T_A)^4$, where T is the stellar effective temperature in the *Kepler* band. The constants A_0, A_{1c}, A_{1s}, and A_{2c} are the phenomenological parameters to describe the phase-curve modulation, and $\phi = 2\pi(t - t_{0,\text{in}})/P_{\text{in}}$ is the orbital phase.[3] These amplitudes, in principle, can be related to the masses of the two bodies with the physical model of ellipsoidal variation and Doppler beaming (Morris and Naftilan 1993; Loeb and Gaudi 2003). We do not use them for the mass estimates, however, because our quarter-by-quarter analysis reveals the temporal variation in the shape of the phase curve. This variation is also consistent with the star-spot modulation nearly synchronized with the orbital motion. Finally, C_{phase} is the overall normalization. In fitting the observed data, $f_{\text{phase}}(t)$ is averaged over 30 min around each time to take into account the long-cadence sampling. The light-travel time effect is neglected in computing $f_{\text{phase}}(t)$ because it is shorter than the data cadence.

As in Sect. 6.2.1, we use an MCMC algorithm to fit the phase-folded light curve for the above parameters. We again include the "jitter" term $\sigma_{\text{LC,phase}}$ in the likelihood $\mathcal{L}_{\text{phase}}$ defined in the same way as in Eq. (6.3). The resulting constraints are in the third column of Table 6.1, and the best-fit light curve is shown in Fig. 6.2b. We also try floating e_{in} and ω_{in}, only to find that the inner orbit is very close to circular. Hence we fix $e_{\text{in}} = 0$ in the following analyses.

The residuals in the bottom panel of Fig. 6.2b exhibit an out-of-eclipse warp and a larger in-eclipse scatter (similar to the one in Bass et al. 2012). The former does not affect our analysis significantly because we do not extract any physical information from the out-of-eclipse modulation. On the other hand, the latter points to systematics that affect the shape of eclipses and thus may bias the resulting system parameters. While it may be due to the spot occultation, ETVs we neglected could also affect the eclipse signal by a similar amount ($A_{\text{ETV}}/(\text{ingress duration}) \sim \mathcal{O}(1\%)$). Although unlikely to explain the random scatter, we also note that the Mandel and Agol (2002) model is exact only for spherical stars and so neglects the tidal distortion of $\mathcal{O}(1\%)$ suggested by the value of A_{2c}. In any case, the results of the following analyses could suffer from that level of systematics, though the main conclusions remain unchanged.

---

[3] Since ETVs we neglected may shift the center of the phase curve, we allow $t_{0,\text{in}}$ used for the phase-curve fitting (denoted as $t_{0,\text{in}}^{\text{phase}}$) to be different from $t_{0,\text{in}}$ in Eq. (6.1). The resulting difference ($|t_{0,\text{in}}^{\text{phase}} - t_{0,\text{in}}| \simeq 5\,\text{min}$) is actually comparable to A_{ETV} and consistent with the ETV origin.

6.3 Geometry and Absolute Dimensions from the Tertiary Event

In this section, we analyze the light curve of the tertiary event jointly with the two components in the previous section. The outer binary motion of the third body is converted to the motions relative to the primary and secondary, which are used to compute their normalized fluxes including the tertiary eclipses, $f_{A,\text{tertiary}}(t)$ and $f_{B,\text{tertiary}}(t)$, with the Mandel and Agol (2002) model. This requires a_{out}/R_A, $\cos i_{\text{out}}$, R_C/R_A, $\Delta\Omega$ (difference in the longitudes of ascending node between inner and outer orbits) and M_B/M_A in addition to the parameters in Sect. 6.2. They are incorporated in the model flux during the tertiary event as

$$f_{\text{tertiary}}(t) = \frac{C_{\text{tertiary}} + \gamma_{\text{tertiary}}(t - t_*)}{1 + F_B/F_A + A_0} \times \left[f_{A,\text{tertiary}}(t) + \frac{F_B}{F_A} f_{B,\text{tertiary}}(t) + A_0 + A_{1c}\cos\phi + A_{1s}\sin\phi + A_{2c}\cos 2\phi \right], \quad (6.5)$$

where C_{tertiary} is the normalization, γ_{tertiary} models the residual instrumental trend around the tertiary event, and we choose $t_*(\text{BJD} - 2454833) = 191.25$. The model likelihood for the tertiary-event light curve $\mathcal{L}_{\text{tertiary}}$ is defined in the same way as in $\mathcal{L}_{\text{phase}}$, again including an additional jitter $\sigma_{\text{LC,tertiary}}$. We first seek for the solution that maximizes $\mathcal{L}_{\text{tertiary}}$ with $\sigma_{\text{LC,tertiary}} = 0$ for various $t_{0,\text{out}}$ using the Levenberg-Marquardt method (Markwardt 2009). Here the above seven new parameters are fitted, while the others are floated within the 3σ boundaries from the ETVs and phase curve (Table 6.1). We then perform an MCMC run from the solution, fitting all the model parameters simultaneously with the joint likelihood $\mathcal{L} \propto \mathcal{L}_{\text{ETV}} \cdot \mathcal{L}_{\text{phase}} \cdot \mathcal{L}_{\text{tertiary}}$. The resulting constraints are summarized in the fourth column of Table 6.1 along with other derived parameters. As shown in Fig. 6.1, our model well reproduces the observed tertiary eclipses. In the following subsections, we discuss the information newly derived from the tertiary eclipses.

6.3.1 Mutual Inclination

Tertiary eclipses on both of the inner two stars suggest a good alignment between inner and outer binary planes. This naive expectation is quantified by our modeling. We obtain $i_{\text{out}} = 89°.83 \pm 0°.02$ and $\Delta\Omega = 3°.2 \pm 0°.6$ (see Fig. 6.3b) as the line-of-sight and sky-plane inclinations of the tertiary orbit. Combined with $i_{\text{in}} = 88°.7 \pm 0°.1$, these results indicate an extremely flat orbital configuration, with the 3σ upper limit on the mutual inclination being $5°$.

6.3.2 Relative Dimensions

Another role of the tertiary event is to determine the mass ratio M_B/M_A and the tertiary-to-primary velocity ratio V_C/V_A during the event, where V is the orbital velocity relative to the center of mass of the inner binary. The constraints are invaluable because they allow us to determine the mass ratios of all three bodies. It is even possible, in principle, to combine them with the ETV amplitude to fix the absolute dimensions of the whole system from photometry alone.

The two quantities, M_B/M_A and V_C/V_A, are closely related to the timings and durations of the three tertiary eclipses. The bottom panel of Fig. 6.3a shows the approximately one-dimensional motion of the inner binary in the sky plane with respect to its center of mass (red and blue sinusoidal lines). Here the motion of the third body (green line) is represented by an almost straight line owing to its long orbital period. For $\Delta\Omega \simeq 0°$, eclipses occur at the intersections of the two lines in this diagram. Thus, the green line should cross either of the red or blue sinusoids at the times of three tertiary eclipses (vertical dashed lines), roughly within the primary/secondary radii (vertical error bars). The condition essentially fixes the amplitude of the blue sinusoid and the slope of the green line, which correspond to M_A/M_B and V_C/V_A, respectively. The ratio V_C/V_A is further constrained by the relative durations of the first and third tertiary eclipses, where the relative velocities between the two stars are $V_A - V_C$ and $V_A + V_C$, respectively.

These ratios yield the relative mass of the third body as well. Using $P_{\rm in}$, $a_{\rm in}/R_A$, $t_{0,\rm out}$, $P_{\rm out}$, $e_{\rm out}$, and $\omega_{\rm out}$ we already derived, V_C/V_A is converted to $a_{\rm out}/R_A$. Since this $a_{\rm out}$ should satisfy Kepler's third law, we obtain

$$\left(\frac{a_{\rm out}/R_A}{a_{\rm in}/R_A}\right)^3 \left(\frac{P_{\rm in}}{P_{\rm out}}\right)^2 = 1 + \frac{M_C/M_A}{1 + M_B/M_A}, \qquad (6.6)$$

which can be solved for M_C/M_A as

$$\frac{M_C}{M_A} = \left[\left(\frac{a_{\rm out}/R_A}{a_{\rm in}/R_A}\right)^3 \left(\frac{P_{\rm in}}{P_{\rm out}}\right)^2 - 1\right]\left(1 + \frac{M_B}{M_A}\right). \qquad (6.7)$$

The mass ratios derived in this way are listed in Table 6.1. These values indicate that the system is dynamically stable, according to the criterion by Mardling and Aarseth (2001).

In fact, the timings of the three eclipses alone allow for other configurations, though they do not fit the eclipse shapes well and hence are rejected (Fig. 6.4).[4] Those in panels (c) and (d) yield too short durations for the third eclipse due to the head-on crossing with one of the inner binary. Moreover, the solutions are unphysical because the values of $a_{\rm out}/R_A$ are so small that $M_C/M_A < 0$ is required in Eq. (6.6).

---

[4] Since these solutions include different M_B/M_A, a radial velocity follow-up is also useful to confirm our solution independently of the possible systematics discussed in Sect. 6.2.2.

6.3 Geometry and Absolute Dimensions from the Tertiary Event

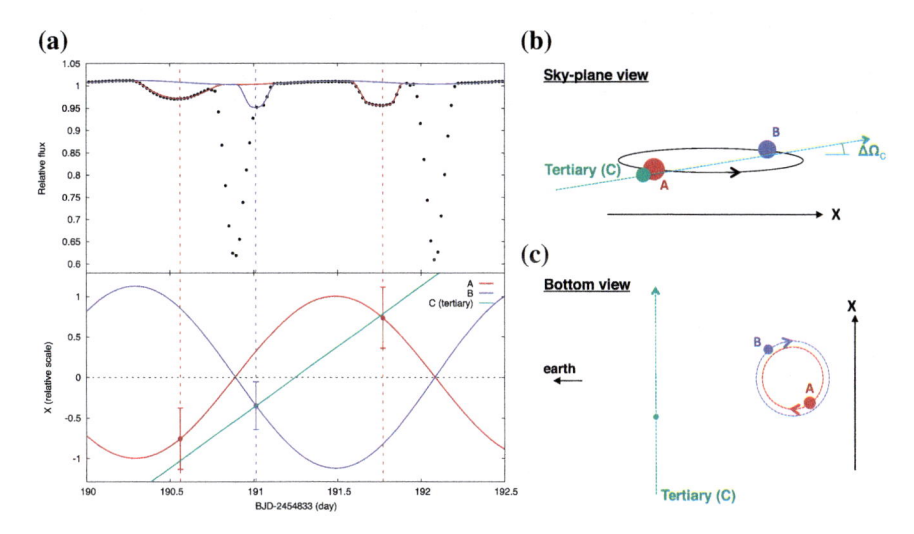

Fig. 6.3 a Relationship between the timings of three tertiary eclipses and motions of three stars. (*Top*) The black dots denote the detrended *Kepler* light curve. The red and blue lines are the best-fit tertiary eclipse models for stars A and B, respectively. The vertical dashed lines denote the rough central times of the tertiary eclipses. (*Bottom*) One-dimensional motion of the three stars (primary: red, secondary: blue, tertiary: green) with respect to the center of mass of the inner binary. The X-axis is defined to coincide with the line of nodes of the inner binary, with its positive direction shown in panels (**b**) and (**c**). The amplitude of the primary motion is normalized to unity, while that of the secondary depends on M_B/M_A (notice that only the relative scale affects the light curve). The vertical bars denote the normalized radii of stars A (red) and B (blue). **b** Sky-plane view and **c** bottom view of the system. Definitions of $\Delta\Omega$ and X-axis are shown schematically

The solution in panel (b), which is the retrograde version of the best solution, fits the light curve better than those in (c) and (d); however, large residuals remain around the first and third tertiary eclipses because R_B is slightly smaller than R_A.

Similarly to F_B/F_A, the constant A_0 could also be related to the third-body temperature by $T_C/T_A = A_0^{1/4}(R_C/R_A)^{-1/2}$, which is also listed in Table 6.1. The value of T_C/T_A thus determined, however, should be considered as a rough upper limit because A_0 includes contaminations from nearby sources and/or systematics in the phase-curve modulation.

6.3.3 Absolute Dimensions

Combined with the ETV amplitude in Eq. (6.2), the mass ratios above can be further used to determine the absolute masses of the system via

$$M_A = 1.074 \times 10^{-3} M_\odot \left(\frac{A_{\rm ETV}}{\rm s}\right)^3 \left(\frac{P_{\rm out}}{\rm day}\right)^{-2} \frac{(1 + M_B/M_A + M_C/M_A)^2}{(M_C/M_A)^3 \sin^3 i_{\rm out}}. \quad (6.8)$$

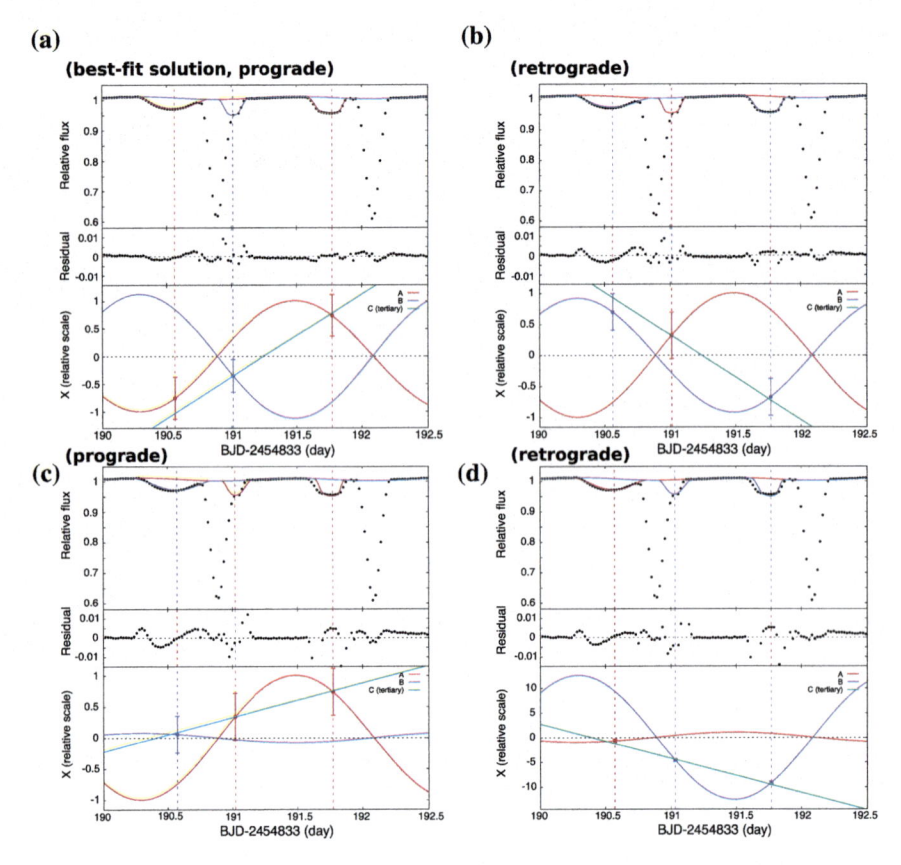

Fig. 6.4 Comparison between the best-fit solution (panel **a**) and other solutions allowed from the timings of the three eclipses alone (panels **b**, **c**, and **d**). The meaning of each panel is basically the same as Fig. 6.3a, but this time the residuals are shown in the middle panels using the same scales

Correspondingly, absolute radii are obtained from $a_{in} = [P_{in}^2 G M_A (1 + M_B/M_A)/4\pi^2]^{1/3}$ and a_{in}/R_A. The constraints on the absolute dimensions, however, are very weak (see Table 6.1) due to the strong correlation $M_A \sim (M_C/M_A)^{-3} \sim (a_{out}/R_A)^{-9}$ as implied by Eqs. (6.7) and (6.8).

The constraints are significantly improved with a better constraint on either M_A or M_C/M_A. To demonstrate this, we repeat the above joint analysis with the Gaussian prior on the primary mass $M_A = 1.15 \pm 0.28\,M_\odot$ based on the value in the *Kepler Input Catalog* (KIC). Here M_A and M_C/M_A are chosen to be fitting parameters instead of a_{out}/R_A and A_{ETV}, where the former two are converted to the latter using Eqs. (6.2) and (6.6). The results are summarized in the last column of Table 6.1, and the parameter correlations for this fiducial solution are illustrated in the joint posterior distribution in Figure C.9. While the constraints on the geometry and relative dimensions are almost unchanged, the absolute masses and radii of all three stars are now determined to the precision similar to the prior constraint. If we also adopt the KIC effective temperature for the primary, we obtain $T_A = T_B = 6100 \pm 200$ K and $T_C < 5000$ K. The dimensions are consistent with the Dartmouth isochrone (Dotter et al. 2008) of ∼7–8 Gyr and suggest that the inner two stars have entered the subgiant branch and that the third body is an M dwarf (Lépine et al. 2013), though the conclusion is sensitive to the priors on M_A and T_A.

6.4 Summary and Discussion

In this chapter, we determine the geometry and physical properties of the hierarchical triple system KIC 6543674 using the *Kepler* photometry alone. Especially, the tertiary event analyzed here enables us to obtain (i) mutual inclination between the inner and outer binary planes, and (ii) mass ratio of the inner binary and instantaneous orbital velocity of the third body. Our analysis clarifies the value of the tertiary eclipses in hierarchical systems with the clear and textbook-like example of the event. The methodology presented here is basically applicable to other hierarchical systems involving tertiary eclipses on both of the inner stars, though more sophisticated models of the eclipse light curve and/or ETVs may be required to accurately model those systems with smaller P_{in} and/or P_{out}/P_{in}. Here it is worth noting that the KIC 6543674 system has the longest P_{out} among the known triply eclipsing hierarchical triples.

The flatness of the system we find (within a few degrees) may have interesting implications for the the origin of the closest binaries, though it is not clear at this point how it compares to the large sample of misaligned triples (Rappaport et al. 2013; Borkovits et al. 2015) as predicted by the KCTF scenario. In this context, a large eccentricity of the third body is intriguing because it may argue for the excitation of the inner orbit's eccentricity by the octupole-order effect e.g., Li et al. 2014 and Petrovich 2015, see also Sect. 3.1.1). In any case, the relative/absolute dimensions of the system as constrained here will be useful for testing those possible alternatives.

Although the absolute dimensions derived above are based on the KIC value, which is of limited reliability, they can be made more accurate with the follow-up spectroscopy to better constrain the stellar photospheric parameters and/or to measure radial velocities, even if they only cover the inner binary orbit. In addition, follow-up photometry of another tertiary event will pin down P_{out} far more precisely, and can also give us some insight into the dynamical interaction in the system. In fact, the non-detection of the second tertiary event in the *Kepler* data, which would have occurred around BJD $= 2456114 \pm 5$ from our result, suggests that the actual period is ∼2σ longer than our estimate and that the second event was hidden in the data gap of about six days centered around BJD $= 2456126$. The fact also motivates the ground-based observation of the next event, which would be around July in 2015.

References

G. Bass, J.A. Orosz, W.F. Welsh et al., ApJ **761**, 157 (2012)
T. Borkovits, S. Rappaport, T. Hajdu, J. Sztakovics, MNRAS **448**, 946 (2015)
J.A. Carter, D.C. Fabrycky, D. Ragozzine et al., Science **331**, 562 (2011)
K.E. Conroy, A. Prša, K.G. Stassun et al., AJ **147**, 45 (2014)
A. Dotter, B. Chaboyer, D. Jevremović et al., ApJS **178**, 89 (2008)
P.P. Eggleton, L. Kiseleva-Eggleton, ApJ **562**, 1012 (2001)
D. Fabrycky, S. Tremaine, ApJ **669**, 1298 (2007)
D. Foreman-Mackey, D.W. Hogg, D. Lang, J. Goodman, PASP **125**, 306 (2013)
T. Hirano, N. Narita, B. Sato et al., ApJ **759**, L36 (2012)
L.G. Kiseleva, P.P. Eggleton, S. Mikkola, MNRAS **300**, 292 (1998)
Y. Kozai, AJ **67**, 591 (1962)
S. Lépine, E.J. Hilton, A.W. Mann et al., AJ **145**, 102 (2013)
G. Li, S. Naoz, B. Kocsis, A. Loeb, ApJ **785**, 116 (2014)
A. Loeb, B.S. Gaudi, ApJ **588**, L117 (2003)
K. Mandel, E. Agol, ApJ **580**, L171 (2002)
R.A. Mardling, S.J. Aarseth, MNRAS **321**, 398 (2001)
C.B. Markwardt, in *Astronomical Society of the Pacific Conference Series, vol. 411, Astronomical Data Analysis Software and Systems XVIII*, ed. by D.A. Bohlender, D. Durand, P. Dowler, p. 251 (2009)
K. Masuda, ApJ **783**, 53 (2014)
K. Masuda, T. Hirano, A. Taruya, M. Nagasawa, Y. Suto, ApJ **778**, 185 (2013)
S.L. Morris, S.A. Naftilan, ApJ **419**, 344 (1993)
J.A. Orosz, ArXiv e-prints, arXiv:1503.07295 (2015)
C. Petrovich, ApJ **805**, 75 (2015)
A. Prša, N. Batalha, R.W. Slawson et al., AJ **141**, 83 (2011)
S. Rappaport, K. Deck, A. Levine et al., ApJ **768**, 33 (2013)
R.W. Slawson, A. Prša, W.F. Welsh et al., AJ **142**, 160 (2011)
B. Thackeray-Lacko, M. Hill, J.A. Orosz et al., in *American Astronomical Society Meeting Abstracts, vol. 221, American Astronomical Society. Meeting Abstracts 221* (142), 09 (2013)

Chapter 7
Summary and Future Prospects

Abstract This chapter summarizes the key results presented in this thesis and discusses possible directions of future studies. We specifically highlight three issues that are important to clarify the role of orbital evolution due to dynamical processes: (i) obliquities of stars hosting long-period transiting planets, (ii) origin of intermediate-period giant planets (warm Jupiters) on eccentric orbits, and (iii) possible difference between single- and multi-transiting systems.

Keywords Warm and cold Jupiters · Single- and multi-transiting systems
Mutual orbital inclination

7.1 Summary

This thesis presented the measurements of stellar obliquities in transiting exoplanetary systems using high-precision photometric data obtained by the *Kepler* space telescope. We also discussed various techniques to determine the architecture of planetary systems not limited to stellar obliquity, which are made possible by, and will expand the potential of, the space-based photometry data. The specific results and achievements in each chapter are summarized as follows.

Chapter 4

- We established a self-consistent methodology to determine the true stellar obliquity by combining asteroseismology, transit light curves, and the RM effect. The methodology was applied for the first time to real systems.
- In the first system, HAT-P-7, the true obliquity of the hot Jupiter host was found to be close to 90°, rather than 180° as implied from the RM measurement. The result relaxes the difficulty in the dynamical origin of this hot Jupiter and its spin–orbit misalignment, because polar orbits are more easily formed than counter-orbiting ones.
- The second system, Kepler-25, hosts two transiting planets with presumably aligned orbits. The equator of the host star was estimated to be slightly tilted with the orbits of two planets, though with a marginal significance, in contrast

to the RM measurements that concluded a spin–orbit alignment. The result, if true, reveals the first spin–orbit misalignment in a multi-planetary system around a main-sequence star, and points to the initial misalignment between the stellar spin and the protoplanetary disk.
- The conclusion for Kepler-25 is not robust at this point, however, as the recent study by Campante et al. (2016) showed that asteroseismology result of this system (and some others) is rather sensitive to the difference in the light-curve processing.

Chapter 5

- We reanalyzed the gravity-darkened transit light curve of Kepler-13A with our own model and obtained an updated solution. We provided a possible solution to the known discrepancy between the gravity-darkening and spectroscopic methods by fully taking into account the uncertainty of the limb-darkening coefficients.
- We modeled the transit shape variation caused by the spin–orbit precession taking advantage of detailed information on the system geometry from gravity-darkened transit model. The model allowed for the empirical determination of the gravitational quadrupole moment of the rotationally deformed host star, and showed that future follow-up observations of λ can be used to test, and even refine, our updated solution.
- We analyzed a similar anomaly in the transit light curve of HAT-P-7b for the first time with the gravity-darkened model. We found a near-polar orbit, independently validating the result in Chap. 4.

Chapter 6

- We showed that three irregular dips in the light curves of the short-period eclipsing binary KIC 6543674 are due to the eclipses caused by the third star gravitationally bound to the inner binary. The orbit of the tertiary star was found to be aligned with the inner one within a few degrees, as expected from the occurrence of such eclipses.
- We combined the modeling of the above tertiary eclipses with the analysis of eclipse timing variations and eclipse light curves of the inner binary to determine the relative and absolute masses and radii of all three stars in the system from the photometric data alone. The system was found to consist of two F sub-giants and outer M dwarf, roughly at the age of 8 Gyr.
- The inferred configuration may be inconsistent with the standard scenario for the close binary formation that involves the Kozai mechanism up to the quadrupole order and tidal dissipation.

The methods of obliquity measurements established in Chaps. 4 and 5 allow us to probe the spin–orbit misalignment of planets with qualitatively different properties than ever explored. They also provide the information complementary to the traditional spectroscopy-based method. The analysis presented in Chap. 6 will be a useful test case to identify and characterize multi-planetary systems that have experienced violent dynamical events and/or experienced the tidal migration following such processes. These efforts will eventually lead to the comprehensive understanding of the origin of the spin–orbit misalignment and its relation to the dynamical history of exoplanetary systems. Indeed, the analyses presented here will also contribute to the future space-based transit surveys planned in the next decade, including *TESS* (Transiting Exoplanet Survey Satellite, Ricker et al. 2014; Sullivan et al. 2015) and *PLATO* (PLAnetary Transits and Oscillations of stars, Rauer et al. 2014).

7.2 Future Prospects—Beyond Hot Jupiters

Originally, the spin–orbit misalignment was a problem specific to hot Jupiters, a rare population of exoplanets. Throughout this thesis, however, we have seen that the problem is beginning to be put in a more general context, motivated by recent obliquity measurements for planets other than hot Jupiters using photometric techniques. This change seems to remind us that the spin–orbit misalignment should eventually be understood coherently with other properties of exoplanetary systems, as an integral part of the comprehensive picture of planet formation and evolution.

We conclude this thesis with presenting possible directions of future studies. In Sect. 7.2.1, we discuss how the methodologies established in this thesis can help to address the key question raised in Chap. 3: is the observed spin–orbit misalignment primordial, or due to dynamical evolution? In Sects. 7.2.2 and 7.2.3, we revisit another problem discussed in Sect. 3.4.2, namely the flatness and mutual orbital misalignment of multi-planetary systems. Here we pursue possible connections between this issue and the stellar obliquity of more generic planetary systems not limited to hot Jupiters, both in terms of characterization of individual systems and analysis of a statistical sample.

7.2.1 Obliquity of Longer-Period Planets Around Hot Stars

To understand whether the high stellar obliquity of hot Jupiters is primordial or acquired, the obliquity measurements of longer-period planets around hot stars (top-right region of Fig. 3.5) will be crucial. If it is of primordial origin, such high obliquities as observed for hot Jupiters around hot stars should occasionally be observed in this region as well. Any difference in the stellar obliquity distributions of hot Jupiters and more distant planets, on the other hand, indicates the important role of dynamical evolution in sculpting the observed obliquity distribution. Note that hot stars are more suited to this sort of inference than cool stars because the close-in planets around hot stars already exhibit high stellar obliquities: the fact implies that any star–planet interaction that damps the stellar obliquity (including tidal dissipation discussed in Sect. 3.2), if present, is less significant for hot stars than cooler ones, for which close-in planets do not frequently exhibit large spin–orbit misalignments.

Currently, this region is almost empty due to the lack of a suitable technique; the obliquity measurement is always challenging for longer-period planets. Moreover, hot stars often have broad spectral lines due to their rapid rotation, or even show no absorption lines at all; these features of hot stars present additional impediments to precise determination of RVs, and hence to the RM measurements. This situation will be improved by systematically applying the gravity-darkening method discussed in Chap. 5 to transiting systems observed with *Kepler* and future space telescopes.

The following estimate shows that the sample of long-period transiting planets around hot stars observed with *Kepler* may already be sufficient to make a meaningful comparison with hot Jupiters. Promising targets of the gravity-darkening analysis should satisfy the following two conditions for a reliable obliquity measurement:

1. the transit signal-to-noise ratio (S/N) is large enough for the anomaly due to gravity darkening, if present, to be detectable, and
2. the host star is not too faint and its rapid rotation can be confirmed with a spectroscopic measurement of $v \sin i_\star$.[1]

We choose S/N > 100 for the first condition, considering that the typical anomaly is $\mathcal{O}(1\%)$ of the transit depth (cf. Sect. 5.1) and that the phase folding increases the transit S/N by at least a factor of a few (i.e., square root of the number of observed transits). The second condition would be satisfied by stars with $K_p < 14.5$, with K_p being the magnitude in the *Kepler* band, which provide the spectrum of roughly S/N ~ 50 for a 30-min exposure with Subaru/HDS. Currently, 98 *Kepler* transiting planet candidates (i.e., KOIs) that fall into the top-right region of Fig. 3.5 (i.e., $T_{\text{eff}} > 6100$ K and $a/R_\star > 10$) satisfy these criteria. On the basis of McQuillan et al. (2014), we estimate the stellar rotation is rapid enough to exhibit significant gravity darkening for about 20% of the sample.[2] Thus, we expect that the stellar obliquity measurement with the gravity-darkening method will be possible for about 20 *Kepler* planets in the top-right region of Fig. 3.5. The value is comparable to the current number ($\gtrsim 30$) of hot Jupiters around hot stars (i.e., top-left region of Fig. 3.5). Also considering the additional sample expected from future surveys, we conclude that the statistical comparison between short- and long-period planets around hot stars using this method could be plausible.

We emphasize that any detection of a large spin–orbit misalignment in the above systematic analysis of long-period planets has a significant implication, if the system exhibits no clear signature of the past dynamical interaction (e.g., orbital planes of multiple planets are well aligned, and/or the orbit is circular; see also Sect. 3.4.1). Such systems, if found, would provide the most direct evidence that the primordial spin–orbit misalignment does exist. They can further be used to estimate the fraction of the primordial spin–orbit misalignment, and thus will play an even more important role if it turns out that the observed spin–orbit misalignment is actually caused by a mixture of the dynamical evolution and initial condition.

---

[1]Note that such measurements are not time critical at all and thus far less demanding than the measurement of λ by observing a spectroscopic transit.

[2]They measured the rotation periods of 34030 main-sequence stars in the *Kepler* field, including 2849 stars with $T_{\text{eff}} > 6100$ K. Among this "hot-star" sample, 569 were found to have rotation periods less than two days; this yields the fraction of 0.2.

7.2 Future Prospects—Beyond Hot Jupiters   127

We note that the same is also true for longer-period planets around cool stars, whose obliquities can be probed with asteroseismology. Here it is interesting to point out that the non-zero eccentricity and the large spin–orbit misalignment are apparently correlated around cool stars (cf. Figs. 3.2 and 3.3), and that no planet on a circular orbit around a cool star has yet been reported to exhibit a significant spin–orbit misalignment. If such a system is identified by asteroseismology, combined with the eccentricity measurement as performed in Sect. 4.4.1, it would also serve as supporting evidence for the primordial misalignment.

7.2.2 Warm Jupiters as Failed Hot Jupiters?

Formation of warm Jupiters is another puzzle similar to the one presented by hot Jupiters, since they are also closer to the host star than theoretically expected. At least several warm Jupiters, such as Kepler-89d (Hirano et al. 2012; Masuda et al. 2013) and Kepler-30d (Sanchis-Ojeda et al. 2012), have other low-mass planets in well-aligned orbits in the same system, and thus are likely to have experienced the gentle disk migration or formed in situ (Huang et al. 2016). On the other hand, there also exists observational evidence suggesting that some warm Jupiters are formed through, or currently experiencing, the high-eccentricity migration as described in Sect. 3.1.

As we mentioned in Sect. 1.1.4, warm Jupiters around metal-rich stars have larger orbital eccentricities than metal-poor counterparts, which suggests the past dynamical interaction (Dawson and Murray-Clay 2013). The same study also showed that three-day pile up (see Sect. 3.1) that seemed absent from the *Kepler* sample is recovered if only the metal-rich sample is considered; this may be a clue that the dynamical interaction enhanced the hot Jupiter formation via the high-eccentricity migration. Dawson and Chiang (2014) then presented a class of warm Jupiter systems whose orbital signatures are consistent with what we expect from the Kozai migration induced by the close companion planet on a misaligned orbit.

Another intriguing feature to note, though its interpretation is rather speculative, is the paucity of warm Jupiters around evolved stars. The trend is clearly shown in Fig. 7.1, in which planets detected with radial velocities (blue circles) are lacking at $a \lesssim 0.5$ AU around stars with radii larger than a few $R_\odot$. Given that many planets with larger semi-major axes are detected, this trend is unlikely to be an observational bias. While the lack of hot Jupiters can be due to the tidal engulfment (e.g., Kunitomo et al. 2011), warm Jupiters seem to be too distant from the host star to be tidally disrupted, unless the efficiency of tidal dissipation increases dramatically as the star evolves (Schlaufman and Winn 2013). This conundrum may be solved if the warm Jupiters are experiencing slow tidal migration due to the Kozai cycle: if their eccentricities are frequently excited to a large value, tidal friction can be significant enough around the pericenter at the high-eccentricity phase to pull the planet into the star (Frewen and Hansen 2016).

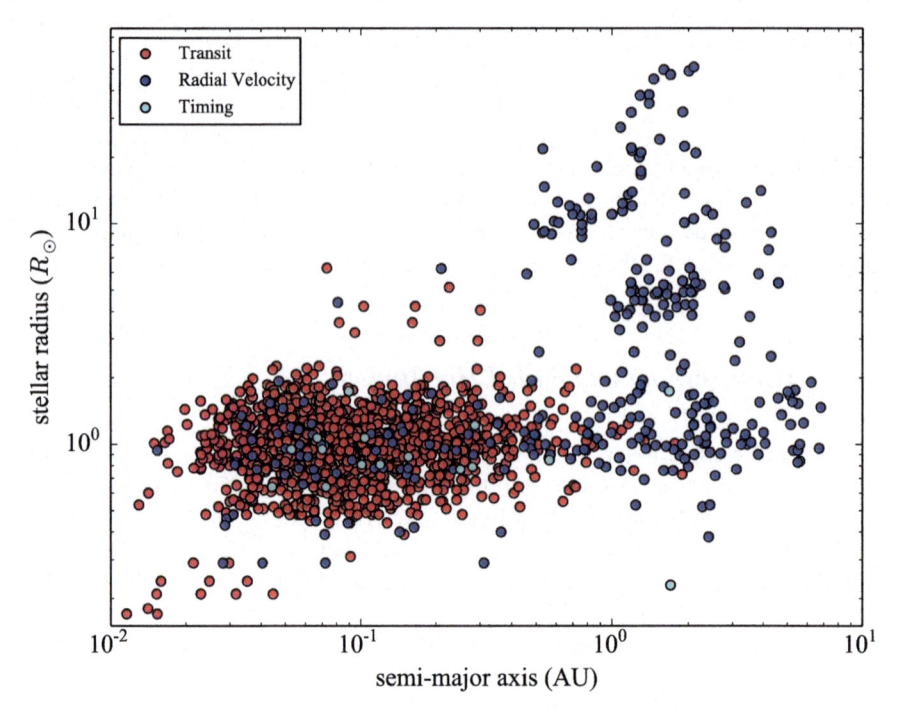

Fig. 7.1 Host star radius versus semi-major axis of known exoplanets. This is based on the same sample as in Fig. 1.1, but stars without radius measurements are excluded

If the intermediate orbits of a warm Jupiter is due to the Kozai migration, it can be shown that the companion must be close, or the orbit rapidly shrinks to become a hot Jupiter as the general relativistic precession terminates the Kozai cycle as the orbit shrinks (Dong et al. 2014). In this case, we expect a similar hierarchical architecture to the one discussed in Chap. 6. Because of their proximity, transiting warm Jupiters may be easier targets to search for such companions responsible for the close-in orbits, than hot Jupiters.

In fact, the systems presented by Dawson and Chiang (2014) are all this type of systems, although they are non-transiting systems confirmed with radial velocities and are not amenable to more detailed characterization (e.g., obliquity or mutual orbital inclination). If the inner warm Jupiter is transiting, we could measure the stellar obliquity and discuss its relationship with the system architecture, and could even constrain the mutual orbital inclination. Such inferences are also applicable to warm Jupiters that will be found more in future transit surveys.

7.2.3 Stellar Obliquity Trend as the Difference in the Planetary System Architecture

Any of the scenarios for the spin–orbit misalignment discussed in Chap. 3, including the primordial ones, attributes the observed λ–T_{eff} trend to the difference in the *stellar* property. This seems to be a reasonable guess in the sense that the steep change in the obliquity distribution around $T_{\text{eff}} = 6100$ K is associated with another property (namely stellar interior structure) that also changes drastically at the same threshold. Nevertheless, it would still be meaningful as well to hypothesize this trend as something associated with the difference in the orbital architecture of *planets*, given the generality of the trend and weak period dependence of obliquities around cool stars (see Sect. 3.2.1).

In this context, the difference in the obliquities in single- and multi-transiting systems may deserve a more in-depth study. Morton and Winn (2014) argued that (some of) single-transiting systems may be parts of multi-planetary systems with large mutual orbital inclinations, unlike the "pancake-flat" multi-planetary systems observed as multi-*transiting* systems. The excess of single-transiting systems in the *Kepler* multiplicity statistics, known as the *Kepler* dichotomy, may also originate from the same "dynamically hot" population as discussed in Sect. 2.4.3.

If the fraction of dynamically hot systems increases with the stellar effective temperature, that may explain both the weak period dependence and generality of the trend resulting from the spot-amplitude analysis (Sect. 2.3.5). While this interpretation is still speculative, Fig. 7.2 shows a suggestive trend: the fraction of multi-transiting systems among the KOI sample starts to decrease for $T_{\text{eff}} \gtrsim 6000$ K. The trend is at least quantitatively consistent with the above speculation, because systems with larger mutual orbital inclinations are less likely to be observed as multi-transiting systems for the same number of planets.

The above result is still tentative in many aspects. For example, the detection bias is not handled very carefully in this analysis; multi-planetary systems may indeed be rare around early-type stars, in which case their decrease is not due to the increasing mutual inclination; or false-positive rates may depend on the stellar mass. A more decisive conclusion will be obtained if the trend is combined with independent information on the number statistics from radial velocity observations, or with different statistics of multi-transiting systems including the period spacing and multiplicity distribution. In any case, this working hypothesis illustrates the possible advantage to study stellar obliquities as an integral part of the planetary system architecture, which should eventually be understood coherently with the physical properties of the planets. Indeed, such a link between the architecture and the physical property has already begun to be pursued (e.g., Dawson et al. 2016).

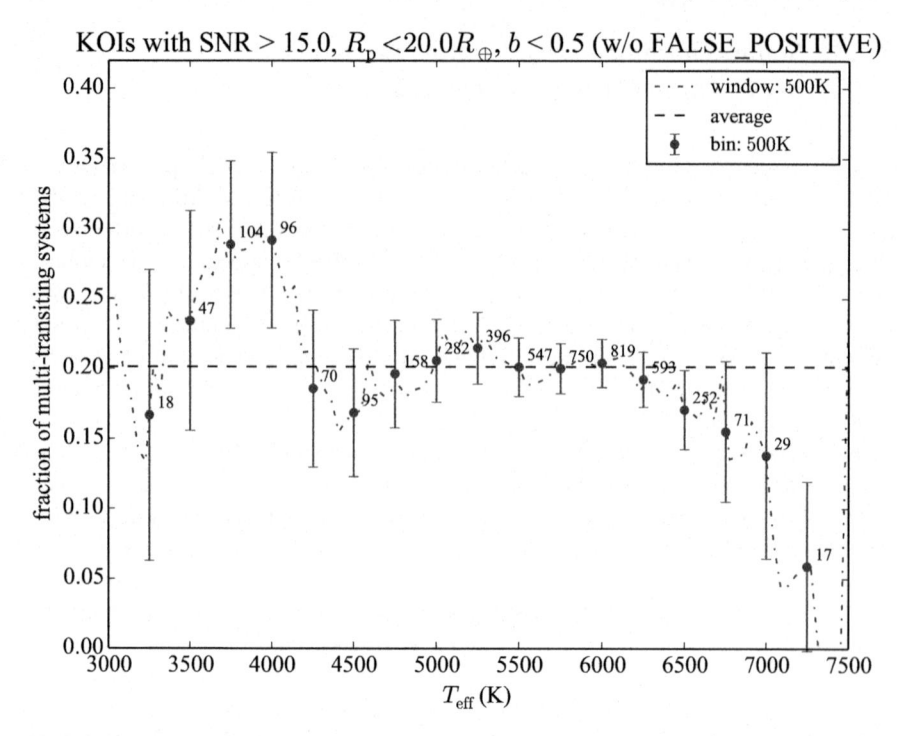

Fig. 7.2 Fraction of multi-transiting KOIs as a function of host-star temperature. The KOIs classified as false positives by the *Kepler* team are all excluded. To attenuate the detection bias against the small planets, the samples are limited to KOIs detected with sufficiently large (>15) signal-to-noise ratios. We also exclude the KOIs with radii larger than $20 R_\oplus$ or with impact parameter b larger than 0.5. The latter condition is to exclude the possible eclipsing binaries, which produce V-shaped eclipses; they usually result in large b when fitted with the transit model. The blue filled circles with error bars are the fractions averaged into 500 K bins, where the number of samples in each bin is shown next to each point and the error bars simply show the Poisson error. The blue dotted line is the running median with the same window size. The black horizontal dashed line shows the average fraction of multi-transiting systems in this sample

References

T.L. Campante, M.N. Lund, J.S. Kuszlewicz et al., ApJ **819**, 85 (2016)
R.I. Dawson, E. Chiang, Science **346**, 212 (2014)
R.I. Dawson, E.J. Lee, E. Chiang, ApJ **822**, 54 (2016)
R.I. Dawson, R.A. Murray-Clay, ApJ **767**, L24 (2013)
S. Dong, B. Katz, A. Socrates, ApJ **781**, L5 (2014)
S.F.N. Frewen, B.M.S. Hansen, MNRAS **455**, 1538 (2016)
T. Hirano, N. Narita, B. Sato et al., ApJ **759**, L36 (2012)
C. Huang, Y. Wu, A.H.M.J. Triaud, ApJ **825**, 98 (2016)
M. Kunitomo, M. Ikoma, B. Sato, Y. Katsuta, S. Ida, ApJ **737**, 66 (2011)
K. Masuda, T. Hirano, A. Taruya, M. Nagasawa, Y. Suto, ApJ **778**, 185 (2013)
A. McQuillan, T. Mazeh, S. Aigrain, ApJS **211**, 24 (2014)
T.D. Morton, J.N. Winn, ApJ **796**, 47 (2014)

References

H. Rauer, C. Catala, C. Aerts et al., Exp. Astron. **38**, 249 (2014)

G.R. Ricker, J.N. Winn, R. Vanderspek et al., Space telescopes and instrumentation 2014: optical, infrared, and millimeter wave, in *Proceedings of SPIE*, vol. 9143 (2014), p. 914320

R. Sanchis-Ojeda, D.C. Fabrycky, J.N. Winn et al., Nature **487**, 449 (2012)

K.C. Schlaufman, J.N. Winn, ApJ **772**, 143 (2013)

P.W. Sullivan, J.N. Winn, Z.K. Berta-Thompson et al., ApJ **809**, 77 (2015)

Appendix A
Planetary Orbit

This appendix summarizes the basic property of the planetary orbit in the two body problem, and specifies the definition of the orbital elements and coordinate system adopted in this thesis.

A.1 The Two-Body Problem

Let us define the orbital elements for the two-body problem under the Newtonian gravity. The equations of motion in this case are

$$m_1 \ddot{\boldsymbol{r}}_1 = +G \frac{m_1 m_2}{|\boldsymbol{r}|^3} \boldsymbol{r}, \tag{A.1}$$

$$m_2 \ddot{\boldsymbol{r}}_2 = -G \frac{m_1 m_2}{|\boldsymbol{r}|^3} \boldsymbol{r}, \tag{A.2}$$

where m_j and $\boldsymbol{r}_j$ are the mass and position vector of the j-th body, G is Newton's gravitational constant, and we define the *relative motion*

$$\boldsymbol{r} = \boldsymbol{r}_2 - \boldsymbol{r}_1. \tag{A.3}$$

The sum of Eqs. (A.1) and (A.2) implies the conservation of the total linear momentum:

$$\boldsymbol{P} \equiv m_1 \dot{\boldsymbol{r}}_1 + m_2 \dot{\boldsymbol{r}}_2 = \text{const}, \tag{A.4}$$

and their difference gives the equation for the relative motion:

$$\ddot{\boldsymbol{r}} = -\frac{GM}{r^3} \boldsymbol{r}, \tag{A.5}$$

© Springer Nature Singapore Pte Ltd. 2018
K. Masuda, *Exploring the Architecture of Transiting Exoplanetary Systems with High-Precision Photometry*, Springer Theses,
https://doi.org/10.1007/978-981-10-8453-9

where $M \equiv m_1 + m_2$ and $r \equiv |\boldsymbol{r}|$. Since the right-hand side of Eq. (A.5) is parallel to $\boldsymbol{r}$, this equation leads to the (specific) angular momentum conservation:

$$\boldsymbol{h} \equiv \boldsymbol{r} \times \dot{\boldsymbol{r}} = \text{const.} \tag{A.6}$$

This means that the relative motion is confined in a plane that is perpendicular to $\boldsymbol{h}$ (orbital plane). Integration of Eq. (A.5) also derives the energy conservation:

$$\frac{1}{2}|\dot{\boldsymbol{r}}|^2 - \frac{GM}{r} = \mathcal{E}, \tag{A.7}$$

where $\mathcal{E}$ is a constant.

A.2 Shape of the Orbit

The trajectory in the orbital plane is obtained by integrating Eq. (A.7) in a polar coordinate system (r, θ). Using $r^2 \dot{\theta} = h \equiv |\boldsymbol{h}|$, Eq. (A.7) reduces to

$$\frac{\dot{r}^2}{2} + \frac{h^2}{2r^2} - \frac{GM}{r} = \mathcal{E}. \tag{A.8}$$

Below we only consider the case of $\mathcal{E} < 0$, i.e., the motion is confined in a finite range of r (bound orbit). Note that $\mathcal{E}$ also has a lower bound, $-\mathcal{E} \leq \frac{1}{2}\left(\frac{GM}{h}\right)^2$, below which no r satisfies Eq. (A.8). Again with $h \equiv |\boldsymbol{h}| = r^2 \dot{\theta}$, Eq. (A.8) is integrated as

$$\theta = \int d\theta = \int \frac{(h/r^2) dr}{\sqrt{2\mathcal{E} + 2GM/r - h^2/r^2}} = \arccos\left(\frac{\frac{h}{r} - \frac{GM}{h}}{\sqrt{2\mathcal{E} + \left(\frac{GM}{h}\right)^2}}\right) + \omega, \tag{A.9}$$

where ω is a constant. Defining

$$a \equiv -\frac{GM}{2\mathcal{E}}, \quad e \equiv \sqrt{1 + \frac{2\mathcal{E} h^2}{G^2 M^2}}, \tag{A.10}$$

Equation (A.9) reduces to

$$r(\theta) = \frac{a(1 - e^2)}{1 + e \cos(\theta - \omega)}. \tag{A.11}$$

Since $0 \leq e < 1$, Eq. (A.11) denotes an ellipse with *semi-major axis a* and *eccentricity e*. The angle ω, the *argument of pericenter*, specifies the point where r becomes minimum (*pericenter* or *periapsis*). So far the reference direction for θ and ω is arbitrary; we will define it in Sect. A.4. The angle $f \equiv \theta - \omega$ is called the *true anomaly*.

Appendix A: Planetary Orbit

A.3 Solution of the Kepler Problem

The motion as a function of time t is also obtained by directly integrating Eq. (A.8), this time without converting dt to $d\theta$:

$$
\begin{aligned}
t = \int dt &= \int \frac{dr}{\sqrt{2\mathcal{E} + 2GM/r - h^2/r^2}} \\
&= \int \frac{dr}{\sqrt{-GM/a + 2GM/r - GMa(1-e^2)/r^2}} \\
&= \sqrt{\frac{a}{GM}} \int \frac{r\, dr}{\sqrt{[ae + (r-a)][ae - (r-a)]}}.
\end{aligned}
\tag{A.12}
$$

Here we introduce the *eccentric anomaly E* via

$$ r - a = -ae \cos E. \tag{A.13} $$

Then Eq. (A.12) can be integrated analytically:

$$ t = \sqrt{\frac{a^3}{GM}} (E - e \sin E) + \tau, \tag{A.14} $$

where τ is a constant.[1] This means that the orbital period P, which is the time for E to increase by 2π, is given by

$$ P = 2\pi \sqrt{\frac{a^3}{GM}}, \quad \text{or} \quad n^2 a^3 = GM, \tag{A.15} $$

where $n \equiv 2\pi/P$ is usually called the *mean motion*. This relation is known as Kepler's third law. Using the mean motion, Eq. (A.14) is rewritten as the Kepler equation,

$$ M = E - e \sin E, \tag{A.16} $$

where

$$ M \equiv n(t - \tau) \tag{A.17} $$

is the *mean anomaly*. Note that τ is the time at which $E = 0$ or $r = a(1 - e)$. Thus τ actually denotes the *time of pericenter passage*.

The above discussion yields a procedure for specifying the motion for a given orbital plane, total mass M, and elements (a, e, ω, τ). At each time t, we compute M using Eqs. (A.15) and (A.17). Then we solve the Kepler equation (A.16) numerically

---

[1] Since $a(1-e) < r < a(1+e)$ from Eq. (A.11), we can define E in the range $[-\pi, \pi)$. Note that r is a decreasing function of E for $E = -\pi \to 0$ (i.e., $\sin E < 0$), while it is increasing for $E = 0 \to \pi$ (i.e., $\sin E > 0$). This distinction is required for correctly integrating Eq. (A.12).

to determine $E(t)$. The eccentric anomaly $E(t)$ is related to r via Eq. (A.13), from which θ or f can also be specified with Eq. (A.11). The explicit relationship between E and f is simply given by

$$\tan\frac{f}{2} = \sqrt{\frac{1+e}{1-e}} \tan\frac{E}{2}. \tag{A.18}$$

A.4 Orientation in Three Dimensions

To completely describe the orbital motion in three dimensions, we also need to specify the direction of the orbital plane, or vector $\boldsymbol{h}$. Throughout the thesis, we adopt right-handed coordinates (XYZ) with $+Z$-axis pointing toward the observer and XY-plane being the sky plane (Fig. A.1).[2] Directions of the XY-axes can be defined arbitrarily. In this coordinate system, we need two angles corresponding to the polar and azimuth angles to describe the direction of $\boldsymbol{h}$. Let us define these angles so that $\hat{\boldsymbol{h}}$, the unit vector of $\boldsymbol{h}$, is given as follows:

$$\hat{\boldsymbol{h}} = \begin{pmatrix} \sin\Omega \sin i \\ -\cos\Omega \sin i \\ \cos i \end{pmatrix}, \tag{A.19}$$

where Ω and i are called the *longitude of the ascending node* and *orbital inclination*, respectively.

The meanings of i and Ω are also illustrated in Fig. A.1. The Z-component of Eq. (A.19) shows that i is the angle between the orbit normal and our line of sight (Z-axis); that is, it is the inclination of the orbital plane with respect to the plane of the sky. The XY-components, on the other hand, indicate that Ω corresponds to the direction of the *ascending node*, where the planet crosses the sky plane with $\dot{Z} > 0$ (i.e., from $Z < 0$ to $Z > 0$), measured from the $+X$-axis. Conventionally, the ascending node is used as a reference direction for θ and ω in Eq. (A.11).

In terms of ω, Ω, and i, the orbit in three dimensions is given by

$$\begin{pmatrix} X \\ Y \\ Z \end{pmatrix} = \begin{pmatrix} \cos\Omega\cos\omega - \sin\Omega\sin\omega\cos i & -\cos\Omega\sin\omega - \sin\Omega\cos\omega\cos i & \sin\Omega\sin i \\ \sin\Omega\cos\omega + \cos\Omega\sin\omega\cos i & -\sin\Omega\sin\omega + \cos\Omega\cos\omega\cos i & -\cos\Omega\sin i \\ \sin\omega\sin i & \cos\omega\sin i & \cos i \end{pmatrix} \begin{pmatrix} r\cos f \\ r\sin f \\ 0 \end{pmatrix}$$

$$= r \begin{pmatrix} \cos\Omega\cos(\omega+f) - \sin\Omega\sin(\omega+f)\cos i \\ \sin\Omega\cos(\omega+f) + \cos\Omega\sin(\omega+f)\cos i \\ \sin(\omega+f)\sin i \end{pmatrix} \equiv \begin{pmatrix} P_X & Q_X & R_X \\ P_Y & Q_Y & R_Y \\ P_Z & Q_Z & R_Z \end{pmatrix} \begin{pmatrix} r\cos f \\ r\sin f \\ 0 \end{pmatrix}.$$

$$\tag{A.20}$$

---

[2] Note that the coordinate system with $+Z$-axis pointing *away* from us is often used as well.

Appendix A: Planetary Orbit

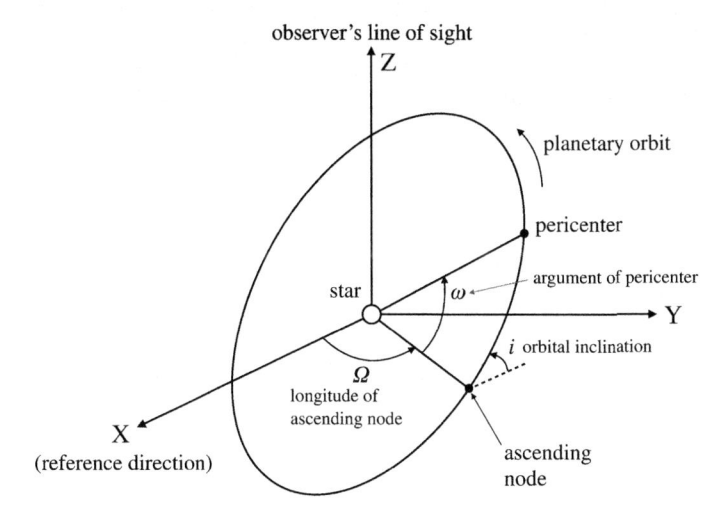

Fig. A.1 Definition of the coordinate system in this thesis and meanings of the longitude of ascending node Ω, orbital inclination i, and argument of pericenter ω

A.5 Summary and Remarks

The meanings of the six orbital elements (upper part of Table A.1) are summarized as follows:

- The *shape* of the orbit is defined by the orbital semi-major axis a and eccentricity e. The former is uniquely related to the orbital energy, while the latter is determined by the energy and angular momentum of the orbital motion.
- The *direction* of the orbit in three dimensions is specified by three angles: argument of pericenter ω, longitude of ascending node Ω, and orbital inclination i. The latter two are essentially the azimuth and polar angles of the angular momentum vector.
- The *position* in the orbit at a given time is specified by the time of pericenter passage τ. Equivalently, we can fix the values of the anomalies at any given time, including M, E, and f.

The anomaly angles defined above are all referred to the pericenter. They are therefore independent of the definition of the coordinate system. In contrast, Ω and i, as well as ω referred to the ascending node, all depend on the specific definition of the coordinate system, or the *reference plane*, to which all these angles are referred. For example, it is often useful to define the ascending node with respect to the invariant plane of the system (plane normal to the total angular momentum), instead of the plane of the sky as we did above. In this case, the ascending node is the intersection between the orbit and the invariant plane (reference plane in this case), and i is the inclination with respect to this plane. For another example, if we define $+Z$-axis away from the observer's direction, as mentioned above, Ω and ω change by π as the ascending and descending nodes are swapped.

Table A.1 Symbols for the orbital elements and their relevant quantities

Symbol	Definition
a	Semi-major axis
e	Eccentricity
i	Inclination
ω	Argument of periastron
Ω	Longitude of ascending node
τ	Time of periastron passage
E	Eccentric anomaly
$\varpi = \omega + \Omega$	Longitude of periastron
f	True anomaly
$\theta = f + \varpi$	True longitude
M	Mean anomaly
$\lambda = M + \varpi$	Mean longitude

Also note that the direction of the pericenter, to which all the anomalies are referred, is not generally constant when the non-Keplerian forces (e.g., general relativity, aspherical star, perturbation from other planets) exist. In this case, the orbital motion may be better represented by the *longitudes*, the angles referred to the axes fixed in an inertial frame.[3] The examples are the *longitude of pericenter* $\varpi \equiv \omega + \Omega$, *mean longitude* $\lambda \equiv M + \varpi$, and *true longitude* $\theta \equiv f + \varpi$. They can also be useful when the orbit is (nearly) circular and the pericenter is not well defined. Even so, the ascending node, and hence longitudes, can always be well defined, unless the orbital plane coincides with the reference plane.

---

[3]Remember that the longitude of ascending node Ω is defined with respect to $+X$-axis; this is why we call Ω a longitude.

Appendix B
Summary of the Transit Method

This appendix provides a more detailed review of the transit method than in Sect. 1.2.3. In observing a transit, we perform a differential photometry: all what we observe is the variation in the *relative* flux of the star as a function of time, and its absolute value does not matter. For this reason, all we can learn from the transit light curve is the geometric properties (i.e., non-dimensional parameters) of a system except for the timescale, from which mean densities of the bodies can be derived. In fact, it is a fairly general conclusion that the mean density is the only dimensional property of a system constrained from the relative flux alone, as long as we consider Newtonian dynamics for point masses.

B.1 Terminology

Following Winn (2010), we define an *eclipse* as the obscuration of one celestial body by another. When the obscuring object is much smaller than the obscured one, this kind of eclipse is called a *transit*, and the opposite case is called an *occultation*. We use the term *grazing* if the obscuration is partial, i.e., the path of a transiting (occulted) object is not totally inside (behind) the larger body. Occultations are often called *secondary eclipses* in exoplanet literatures.

B.2 Transit Geometry

Equation (A.20) gives the sky-projected star–planet distance $r_{\rm sky} \equiv \sqrt{X^2 + Y^2}$ as

$$r_{\rm sky} = \frac{a(1-e^2)}{1+e\cos f}\sqrt{1-\sin^2(\omega+f)\sin^2 i}. \tag{B.1}$$

© Springer Nature Singapore Pte Ltd. 2018
K. Masuda, *Exploring the Architecture of Transiting Exoplanetary Systems with High-Precision Photometry*, Springer Theses, https://doi.org/10.1007/978-981-10-8453-9

Note that r_{sky} does not depend on Ω due to the rotational symmetry of the system with respect to the line of sight.

The planetary transit, if visible, is centered on the minimum value of this r_{sky}, which we define as the *transit center*. The value of f that minimizes r_{sky} is obtained by solving $dr_{sky}/df = 0$, and this equation reduces to

$$\Delta = \frac{1}{2} \arcsin\left[2e \cos(\omega + \Delta) \left(\frac{1}{\sin^2 i} - \cos^2 \Delta\right) - e \sin(\omega + \Delta) \sin 2\Delta\right], \tag{B.2}$$

where we define $\Delta \equiv \pi/2 - (\omega + f)$. This can be solved by iteration to give $\Delta = e \cos \omega \cot^2 i - e^2 \sin 2\omega \cot^2 i (1 + \cot^2 i) + \mathcal{O}(e^3)$, which is negligibly small except for planets on highly eccentric (e is large) and close-in orbits with grazing eclipses (i is far from $\pi/2$). The true anomaly at the transit center, therefore, is well approximated by

$$f_{tra} = +\frac{\pi}{2} - \omega, \tag{B.3}$$

i.e., transits are centered at *inferior conjunctions*. In this approximation, the star–planet distance in the sky plane at the transit center is given by

$$r_{sky}\left(f = +\frac{\pi}{2} - \omega\right) = a \cos i \frac{1 - e^2}{1 + e \sin \omega} \equiv bR_\star, \tag{B.4}$$

where we define the normalized *impact parameter* of the transit, b, in the last equality.

Using the impact parameter defined above, the condition for the transit to be observable at all for a given observer is written as

$$b < \frac{R_\star + R_p}{R_\star} \quad \text{or} \quad |\cos i| < \frac{R_\star + R_p}{a} \frac{1 + e \sin \omega}{1 - e^2} \equiv \cos i_0. \tag{B.5}$$

Thus, the transit probability for a randomly placed observer is

$$p_{tra} = \frac{\int_{-\cos i_0}^{\cos i_0} d\cos i}{\int_{-1}^{+1} d\cos i} = \cos i_0 = \frac{R_\star + R_p}{a} \frac{1 + e \sin \omega}{1 - e^2}. \tag{B.6}$$

Note that the measure $d\cos i$ comes from the inclination dependence of the solid angle (proportional to $\sin i$). If ω is not known either, we also average over ω to obtain

$$p_{tra} = \frac{R_\star + R_p}{a} \frac{1}{1 - e^2} \simeq 0.005 \left(\frac{R_\star}{R_\odot}\right) \left(\frac{a}{1 \text{ AU}}\right)^{-1} \frac{1}{1 - e^2}. \tag{B.7}$$

The corresponding formulae for the occultation can be derived in an analogous manner. The true anomaly at the occultation center is replaced by $-\pi/2 - \omega$ in Eq. (B.3), and so the signs of $e \sin \omega$ are all flipped in Eqs. (B.4) through (B.6) for the occultation case. Equation (B.7) remains the same.

Appendix B: Summary of the Transit Method

B.3 Information from Eclipses

If the transit is observed, its depth reveals the planet-to-star radius ratio, which is never constrained by other methods. In addition, the timings of the repeating transits allow us to constrain the orbital phase of the planet, while the transit shape yields the geometric parameters of the orbit including *scaled semi-major axis* $a/R_\star$ and orbital inclination i. We will see this using a simplified model of the transit in Appendix B.3.1. We will also comment on how the mean stellar density and, in some cases, orbital eccentricity can be derived from the time-domain information of the light curve.

B.3.1 Constraints on Geometry from the Transit Shape

Here we describe the relationship between the shape of the transit and geometrical parameters of the system. We adopt a simple "trapezoidal" model, where the shape of the transit light curve is approximated by a trapezoid. In fact, the simple model is enough to capture the essential property of the light curve, and the parameters that can be constrained from the modeling is basically the same as obtained from a more elaborate model (e.g., Mandel and Agol 2002, used in Chaps. 4 through 6).

B.3.1.1 Circular Orbit

Neglecting the effect of the stellar limb darkening, the shape of the extinction due to a planetary transit is well approximated by a simple trapezoid as shown in Fig. B.1. In this case, the shape of the light curve is characterized by

1. the *transit depth*: $\delta \equiv$ (relative decrease in the stellar flux),
2. the *total duration of the transit*: $T_{\rm tot} \equiv t_{\rm IV} - t_{\rm I}$,
3. the *duration of the full transit*: $T_{\rm full} \equiv t_{\rm III} - t_{\rm II}$,

where the durations $T_{\rm tot}$ and $T_{\rm full}$ are defined through the four contact times illustrated in Fig. B.1. We also define the durations of ingress and egress, $\tau_{\rm ing} = t_{\rm II} - t_{\rm I}$ and $\tau_{\rm egr} = t_{\rm IV} - t_{\rm III}$. When the orbit is circular, $\tau_{\rm ing}$ and $\tau_{\rm egr}$ are equal and related to the above durations as $\tau \equiv \tau_{\rm ing} = \tau_{\rm egr} = (T_{\rm tot} - T_{\rm full})/2$.

These parameters are simply related to the geometric parameters of a planet and its orbit in the following manner. The transit depth δ is given as the fraction of the stellar flux blocked by the planet to the whole stellar flux:

$$\delta = \left(\frac{R_{\rm p}}{R_\star}\right)^2. \tag{B.8}$$

Fig. B.1 Illustration of the transit (*upper panel*) and the in-transit flux approximated as a trapezoid (*lower panel*). Four contact times are defined

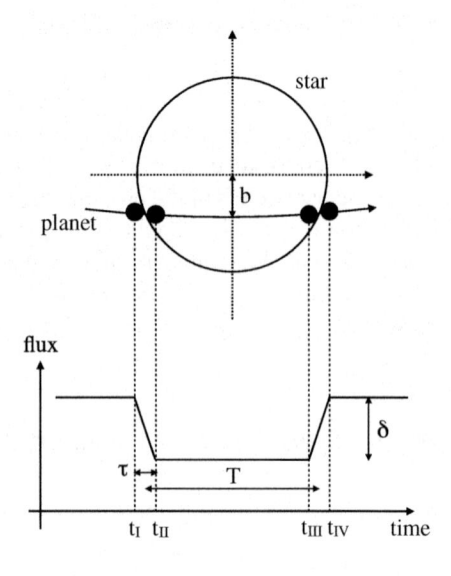

The angle the planet needs to travel during a transit, divided by its angular velocity, yields the two durations as

$$T_{\text{tot}} = \frac{P}{\pi} \sin^{-1} \left[\frac{R_\star \sqrt{(1 + R_p/R_\star)^2 - b^2}}{a \sin i} \right], \tag{B.9}$$

$$T_{\text{full}} = \frac{P}{\pi} \sin^{-1} \left[\frac{R_\star \sqrt{(1 - R_p/R_\star)^2 - b^2}}{a \sin i} \right]. \tag{B.10}$$

In the limiting case that $R_p/R_\star \ll 1$ and $R_\star/a \ll 1$, these results are greatly simplified:

$$T \equiv \frac{T_{\text{tot}} + T_{\text{full}}}{2} \simeq T_{\text{tot}} \simeq T_{\text{full}} \simeq T_0 \sqrt{1 - b^2}, \tag{B.11}$$

$$\tau \simeq \frac{T_0}{\sqrt{1 - b^2}} \frac{R_p}{R_\star}, \tag{B.12}$$

where T_0 is a characteristic timescale given by

$$T_0 \equiv \frac{R_\star P}{\pi a} \simeq 13\,\text{h} \left(\frac{P}{1\,\text{yr}} \right)^{1/3} \left(\frac{\rho_\star}{\rho_\odot} \right)^{-1/3}. \tag{B.13}$$

Appendix B: Summary of the Transit Method

The above expressions for $(\delta, T_{\text{tot}}, T_{\text{full}})$ can be inverted to give a set of geometrical parameters

$$\frac{R_p}{R_\star} = \sqrt{\delta}, \tag{B.14}$$

$$b^2 = \left(\frac{a}{R_\star}\cos i\right)^2 = \frac{(1-\sqrt{\delta})^2 - (T_{\text{full}}/T_{\text{tot}})^2(1+\sqrt{\delta})^2}{1-(T_{\text{full}}/T_{\text{tot}})^2} \simeq 1 - \frac{T}{\tau}\sqrt{\delta}, \tag{B.15}$$

$$\frac{R_\star}{a} = \frac{\pi}{2\delta^{1/4}}\frac{\sqrt{T_{\text{tot}}^2 - T_{\text{full}}^2}}{P} \simeq \frac{\pi}{\delta^{1/4}}\frac{\sqrt{\tau T}}{P}, \tag{B.16}$$

where the last approximation holds when $\tau \ll T$. In this way, the dimensionless parameters that characterize the transit shape, $(\delta, T/P, \tau/P)$, are related to the planetary radius R_p and two orbital elements a and i in units of the stellar radius $R_\star$ for lengths. The same is true even when we use more detailed transit models such as Mandel and Agol (2002).

B.3.1.2 Eccentric Orbit

If the orbit is eccentric, the durations (B.9) and (B.10) are calculated via

$$t_\beta - t_\alpha = \int_{t_\alpha}^{t_\beta} dt = \int_{f_\alpha}^{f_\beta}\left(\frac{df}{dt}\right)^{-1} df = \frac{P(1-e^2)^{3/2}}{2\pi}\int_{f_\alpha}^{f_\beta}\frac{1}{(1+e\cos f)^2} df, \tag{B.17}$$

where $\alpha, \beta = $ I, II, III, IV. Here we use $r^2 \dot{f} = h = na^2\sqrt{1-e^2}$ and Eq. (A.11) in the last equality, and f_α is the solution of

$$r_{\text{sky}}(f_\alpha) = \frac{a(1-e^2)}{1+e\cos f_\alpha}\sqrt{1-\sin^2(\omega+f_\alpha)\sin^2 i} = R_\star \pm R_p, \tag{B.18}$$

where $+$ and $-$ signs correspond to $\alpha = $ I, IV and $\alpha = $ II, III, respectively. Equation (B.18) cannot be solved analytically for f_α, but the solution to the leading orders of e and $R_\star/a$ can be obtained as

$$\frac{\pi}{2} - (\omega + f_\alpha) = \frac{1}{\sin i}\frac{R_\star}{a}\frac{1+e\sin\omega}{1-e^2}\sqrt{\left(1\pm\frac{R_p}{R_\star}\right)^2 - b^2} \tag{B.19}$$

for $\alpha = $ I, II. As the first approximation, therefore, T_{tot} and T_{full} for the eccentric case are

$$T_{\text{tot}} = \frac{P(1-e^2)^{3/2}}{2\pi} \cdot 2\left[\frac{\pi}{2} - (\omega + f_1)\right] \frac{1}{(1+e\sin\omega)^2}$$

$$= \frac{P}{\pi} \frac{R_\star}{a} \frac{\sqrt{(1+R_p/R_\star)^2 - b^2}}{\sin i} \left(\frac{\sqrt{1-e^2}}{1+e\sin\omega}\right), \quad \text{(B.20)}$$

$$T_{\text{full}} = \frac{P}{\pi} \frac{R_\star}{a} \frac{\sqrt{(1-R_p/R_\star)^2 - b^2}}{\sin i} \left(\frac{\sqrt{1-e^2}}{1+e\sin\omega}\right). \quad \text{(B.21)}$$

These are different from the circular case by the factor in the parentheses, corresponding to the difference of orbital velocity around the transit.

Since the non-zero eccentricity also introduces the velocity asymmetry with respect to the transit center, τ_{ing} and τ_{egr} are generally unequal. The difference is, however, usually negligibly small. To the leading orders of $R_\star/a$ and e, we have

$$\frac{\tau_{\text{egr}} - \tau_{\text{ing}}}{\tau_{\text{egr}} + \tau_{\text{ing}}} \sim e\cos\omega \left(\frac{R_\star}{a}\right)^3 (1-b^2)^{3/2} \quad \text{(B.22)}$$

(Winn, 2010). This quantity is less than $10^{-2}e$ for a close-in planet with $R_\star/a = 0.2$, and even smaller for more distant planets.

Considering the possible non-zero eccentricity, therefore, only affects the estimate of $R_\star/a$, while $R_p/R_\star$ and b are unchanged. The modified formula for $R_\star/a$ is

$$\frac{R_\star}{a} = \frac{\pi}{2\delta^{1/4}} \frac{\sqrt{T_{\text{tot}}^2 - T_{\text{full}}^2}}{P} \frac{1+e\sin\omega}{\sqrt{1-e^2}} \equiv \left(\frac{R_\star}{a}\right)_{\text{circ}} \frac{1+e\sin\omega}{\sqrt{1-e^2}} \simeq \frac{\pi}{\delta^{1/4}} \frac{\sqrt{\tau T}}{P} \frac{1+e\sin\omega}{\sqrt{1-e^2}}, \quad \text{(B.23)}$$

where $(R_\star/a)_{\text{circ}}$ is $R_\star/a$ derived assuming $e=0$. Correspondingly, the inclination estimated from the transit shape is also modified as

$$\cos i = b \frac{R_\star}{a} \frac{1+e\sin\omega}{1-e^2} \equiv \cos i_{\text{circ}} \frac{(1+e\sin\omega)^2}{(1-e^2)^{3/2}} \simeq \left[1 - \frac{T}{\tau}\sqrt{\delta}\right]^{1/2} \frac{\pi}{\delta^{1/4}} \frac{\sqrt{\tau T}}{P} \frac{(1+e\sin\omega)^2}{(1-e^2)^{3/2}}, \quad \text{(B.24)}$$

where $\cos i_{\text{circ}}$ is $\cos i$ derived assuming $e=0$. This means that the eccentricity cannot be constrained from the transit shape alone without an independent constraint on $R_\star/a$.[4]

B.3.1.3 Constraint on the Phase

Transit observations constrain the orbital phase of the planet, in addition to the geometric parameters discussed above.

---

[4] In principle, the prior constraint on i could also be useful. However, it is usually impossible to constrain i with a sufficient precision, independently from the transit.

Appendix B: Summary of the Transit Method

First let us consider the case of a circular orbit. Assuming that the transits are observed repeatedly, the series of observed transit times fix the orbital period P and the time of a transit center t_0 (sometimes called *transit epoch*). For $e = 0$, Eqs. (A.16) and (A.18) yield $M = E = f$, and now $\omega + f(t_0) = \pi/2$ from Eq. (B.3). Thus, $\omega + f$ (i.e., orbital phase) at any time t is completely specified with t_0 and P:

$$\omega + f(t) = \frac{\pi}{2} + \frac{2\pi}{P}(t - t_0). \tag{B.25}$$

This is equivalent to fixing τ, although it is not uniquely defined for a circular orbit.

For $e \neq 0$, we use Eq. (A.16) to obtain

$$\tau = t_0 - \frac{E(t_0) - e \sin E(t_0)}{n}, \tag{B.26}$$

where $E(t_0)$ is derived from Eqs. (B.3) and (A.18):

$$E(t_0) = 2 \arctan\left[\sqrt{\frac{1-e}{1+e}} \tan\left(\frac{\pi}{4} - \omega\right)\right]. \tag{B.27}$$

B.3.2 Constraint on the Physical Dimension

Dividing Kepler's third law (A.15) by $R_\star^3$, we obtain

$$\left(\frac{2\pi}{P}\right)^2 \left(\frac{a}{R_\star}\right)^3 = \frac{GM}{R_\star^3} = \frac{4\pi G}{3}\rho_\star\left(1 + \frac{M_p}{M_\star}\right), \tag{B.28}$$

where $\rho_\star$ is the *mean stellar density*. Thus, neglecting $M_p/M_\star \lesssim 10^{-3}$ for a star–planet system, the mean stellar density is obtained purely from the transit observables (Seager and Mallén-Ornelas 2003), as long as the eccentricity is already constrained.

Notice that $R_p/R_\star$, $a/R_\star$, and i derived in Appendix B.3.1 are determined solely by the *dimensionless* shape parameters of the transit, δ, T/P, and τ/P, and that the information on the *absolute* timescale (i.e., P) is used for the first time in Eq. (B.28). As we will discuss in more detail below, $\rho_\star$ (or mean density in general) is the only dimensional quantity constrained from the light curve alone.

B.3.3 Constraints on Orbital Eccentricity

It is generally difficult to constrain the orbital eccentricity from the transit light curve alone. Nevertheless, the orbital eccentricity is (partly) determined from the light

B.3.3.1 Timing and Duration of the Occultation

If the occultation (secondary eclipse) is observed for a close-in planet, the orbital eccentricity is fully constrained from the light curve from the timing and duration of the occultation relative to the transit. In this case, the mean stellar density is derived from the light curve without ambiguity via Eqs. (B.28) and (B.23).

First, the time from the transit to the occultation, $\Delta t_{\text{tra}\to\text{occ}}$, is computed in a similar manner to Eq. (B.17):

$$\Delta t_{\text{tra}\to\text{occ}} = \int_{f_{\text{tra}}}^{f_{\text{occ}}} \left(\frac{df}{dt}\right)^{-1} df = \frac{P(1-e^2)^{3/2}}{2\pi} \int_{3\pi/2-\omega}^{-\pi/2-\omega} \frac{df}{(1+e\cos f)^2},$$

$$\simeq \frac{P(1-e^2)^{3/2}}{2\pi}(\pi + 4e\cos\omega) = \frac{P}{2}\left(1 + \frac{4}{\pi}e\cos\omega\right) + \mathcal{O}(e^2). \quad (\text{B.29})$$

This means that $\Delta t_{\text{tra}\to\text{occ}}$ deviates from $P/2$ due to the asymmetry of the orbit with respect to the line of sight, which is represented by $e\cos\omega$.[5] Second, Eqs. (B.21) and (B.20) show that durations relevant to the transit and occultation, T_{tra} and T_{occ}, are related by

$$\frac{T_{\text{tra}}}{T_{\text{occ}}} = \frac{1 - e\sin\omega}{1 + e\sin\omega}, \quad (\text{B.30})$$

because the corresponding formulae for the occultation is obtained by replacing $e\sin\omega$ with $-e\sin\omega$.[6] The difference in the durations comes from that in the velocities. Therefore the duration ratio is sensitive to the orbit asymmetry with respect to the sky plane, which is represented by $e\sin\omega$. Note that $e\cos\omega$ is more precisely constrained than $e\sin\omega$ because $\Delta t_{\text{tra}\to\text{occ}}$ is longer than $T_{\text{tra/occ}}$ roughly by a factor of $a/R_\star$. These two effects are schematically illustrated in Fig. B.2.

B.3.3.2 Constraint on the Mean Stellar Density

As discussed in Sect. 4.3, asteroseismology can precisely constrain the mean stellar density $\rho_\star$. Less precise constraints on $\rho_\star$ can also be derived from the spectroscopic observation. Such a constraint on $\rho_\star$ allows us to determine $a/R_\star$ via Eq. (B.28) independently from the transit light curve. This $a/R_\star$ can be combined with Eq. (B.23) to

---

[5] Remember that ω is measured from the sky plane, and so the major-axis of the orbit coincides with the line of sight for $\omega = \pm\pi/2$.

[6] Note that the formulae for the occultation is the same as the transit formulae in the coordinate system reflected with respect to the plane of the sky. As we noted in Appendix A.5, ω and Ω changes by π in this case.

Appendix B: Summary of the Transit Method

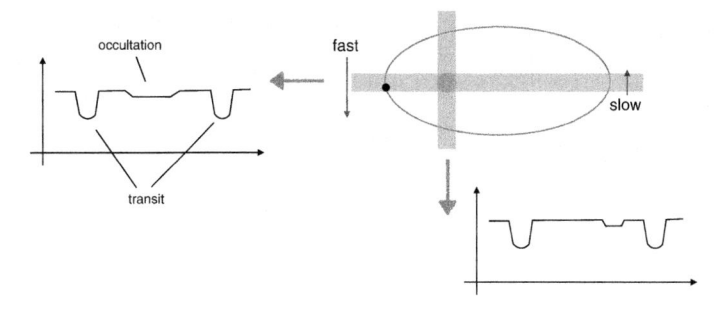

Fig. B.2 Schematic illustration of the effect of a non-zero eccentricity on the transit and occultation light curves. When the system is observed from the direction of the left orange arrow, the occultation is elongated due to a positive $e \sin \omega$. When the observer is in the direction of the bottom arrow, on the other hand, the occultation occurs later than $P/2$ because $e \cos \omega > 0$

contrain $(1 + e \sin \omega)/\sqrt{1 - e^2}$. Since $(1 + e \sin \omega)/\sqrt{1 - e^2} \leq \sqrt{(1 + e)/(1 - e)}$, the method (only) gives a lower limit on e (see, e.g., Van Eylen & Albrecht (2015) and Uehara et al. (2016) for the application).

B.4 Remarks on the Similarity

Why is the transit light curve related to the mean stellar density, while the other parameters are only obtained in a non-dimensional manner? This property is essentially due to the similarity of the problem and hence applies fairly generally.

B.4.1 Stellar Intensity Profile

First, we note that the light curve is only sensitive to the planetary orbit normalized to the stellar radius $R_\star$ as long as the following conditions are satisfied.

Suppose that the stellar intensity profile depends on the position only through $r_\star/R_\star$, where $r_\star$ is a two-dimensional vector specifying a position on the stellar disk. We write such a profile as $I(r_\star/R_\star; \alpha_\star)$, where $\alpha_\star$ denotes the set of parameters that describes the stellar intensity profile, and suppose that $I = 0$ outside of the stellar disk. In addition, we assume that the planetary disk modifies the stellar intensity by a factor of $E([r_\star - r_p]/R_\star; \beta_\star)$, where r_p here is the position of the planet center on the sky plane, and $\beta_\star$ is the set of parameters that specify the shape and intensity profile of the planetary disk. Then, the relative flux f, as an observable of the differential photometry, is solely determined by $\tilde{r}_p(t) \equiv r_p(t)/R_\star$:

$$f = \frac{\int_{I(r'_\star/R_\star;\alpha_\star)\neq 0} E([r'_\star - r_p]/R_\star; \beta_p) I(r'_\star/R_\star; \alpha_\star) d^2 r'_\star}{\int_{I(r'_\star/R_\star;\alpha_\star)\neq 0} I(r'_\star/R_\star; \alpha_\star) d^2 r'_\star}$$

$$= \frac{\int_{I(\tilde{r}'_\star;\alpha_\star)\neq 0} E(\tilde{r}'_\star - \tilde{r}_p; \beta_p) I(\tilde{r}'_\star; \alpha_\star) d^2 \tilde{r}'_\star}{\int_{I(\tilde{r}'_\star;\alpha_\star)\neq 0} I(\tilde{r}'_\star; \alpha_\star) d^2 \tilde{r}'_\star}$$

$$= f(\tilde{r}_p(t); \alpha_\star, \beta_p). \tag{B.31}$$

The above conclusion applies to broad situations. For example, the standard transit model used in Chaps. 4 through 6 assumes the quadratic limb-darkening law and dark planetary disk:

$$I(r_\star/R_\star; (u_1, u_2)) = \begin{cases} I(0)\left[1 - u_1(1-\mu) - u_2(1-\mu)^2\right], & \mu = \sqrt{1-(r_\star/R_\star)^2} \text{ for } r_\star/R_\star < 1 \\ 0 & \text{otherwise} \end{cases} \tag{B.32}$$

$$E([r_\star - r_p]/R_\star; R_p/R_\star) = \begin{cases} 0 & \text{for } |r_\star - r_p|/R_\star < R_p/R_\star \\ 1 & \text{otherwise} \end{cases} \tag{B.33}$$

where u_1 and u_2 are constants called *limb-darkening coefficients*. While $\tilde{r}_p$ depends on $(P, a/R_\star, e, \omega, i, \Omega, \tau)$, the profile (B.32) is axisymmetric and so f does not depend on Ω. Therefore, $f = f(\tilde{r}_p(t); (u_1, u_2), R_p/R_\star)$ is sensitive to $(P, a/R_\star, e, \omega, i, \tau, u_1, u_2, R_p/R_\star)$; this is indeed the set of parameters that can in principle be constrained from the light curve, as shown in Appendix B.3.

The same property also applies to the gravity-darkened transit model discussed in Chap. 5. As described in detail in Sect. 5.2.1, the stellar intensity profile in this case is solely determined by the temperature at the stellar pole $T_{\star,\text{pole}}$, gravity-darkening exponent β, and the direction of the surface gravity vector at each point on the stellar surface. Since the last one depends on the radius vector normalized by $R_\star$, direction of the stellar spin axis, and a dimensionless parameter $\gamma = 3\pi f_{\text{rot}}^2/2G\rho_\star$ with f_{rot} being the stellar rotation frequency, the intensity profile has the form of $I(r_\star/R_\star; (\rho_\star, f_{\text{rot}}, T_{\star,\text{pole}}, \beta, i_\star, \Omega_\star))$, where $\Omega_\star$ is defined analogously to Eq. (A.19) for the stellar spin vector. This explains the choice of the model parameters in Sect. 5.2.1. Note that only $\lambda = \Omega - \Omega_\star$ is constrained due to the arbitrariness in choosing the reference direction ($+X$-axis), and that $M_\star$ is not included in the light curve model explicitly but only constrained through that on $v \sin i_\star$ in Eq. (5.3).

B.4.2 Newtonian Gravity

We have seen in Appendix B.3.2 that $a/R_\star$ and P obtained from the transit light curve can be combined to determine the mean stellar density $\rho_\star$, if we neglect $M_p/M_\star \ll 1$. Why does the light curve give $\rho_\star$ as the only absolute dimension of the system? This is because the timescale of the motion, which is the only dimensional quantity obtained from the light curve, is determined by the density in the Newtonian gravity.

Appendix B: Summary of the Transit Method

The dimensionless form of Eq. (A.5) explicitly shows this property:

$$\frac{d^2(\boldsymbol{r}/R_\star)}{d(t/\sqrt{R_\star^3/GM_\star})^2} = -\frac{\boldsymbol{r}/R_\star}{(r/R_\star)^3}, \tag{B.34}$$

which means that the normalized motion $\boldsymbol{r}/R_\star$ is a function of t/t_{ff}, where

$$t_{\mathrm{ff}} \equiv \sqrt{\frac{3}{4\pi G \rho_\star}}. \tag{B.35}$$

This property, along with the fact that the light curve only depends on $r_\mathrm{p}/R_\star$, leads to the following conclusion: the observed transit light curve is invariant under any scaling of the mass and radius that keeps the mean stellar density unchanged. In other words, the transit light curve is only sensitive to the relative dimensions of the system except for the mean stellar density. Chapter 6 deals with an exception to this rule, where the timescale is related to another absolute dimension (i.e., size) of the system via the speed of light c. Even in this case, the absolute mass is not well determined because the system size depends only on the cubit root of the mass scale.

B.4.2.1 TTVs Constrain the Mean Densities of the Bodies Alone

Since the above property comes from the scaling of the Newtonian dynamics, it can be generalized to the transit light curve of multi-planetary systems, where the member planets gravitationally perturb each other to produce the deviation from the two-body case (e.g., transit timing variations or TTVs; see also Sect. 1.2.3). In this case, the equations of motion for the coordinates normalized by $R_\star$ depend on masses of the planets divided by $M_\star$, as well as $\rho_\star$. For transiting planets, we can also constrain $R_\mathrm{p}/R_\star$. In the ideal situation where the member planets are strongly interacting and all transiting, therefore, TTVs allow for constraining the planetary density $\rho_\mathrm{p} = \rho_\star (M_\mathrm{p}/M_\star)(R_\mathrm{p}/R_\star)^{-3}$ purely from the light curve. On the other hand, the absolute mass and radius scales are never constrained from the light curve alone, as the same scaling property as discussed above holds even in the presence of planet–planet interaction. Note that this is true even when the transit variations other than TTVs (e.g., transit duration variations) are considered, as pointed out by Sanchis-Ojeda et al. (2012).

References

K. Mandel, E. Agol, ApJ **580**, L171 (2002)
R. Sanchis-Ojeda, D.C. Fabrycky, J.N. Winn et al., Nature **487**, 449 (2012)
S. Seager, G. Mallén-Ornelas, ApJ **585**, 1038 (2003)
S. Uehara, H. Kawahara, K. Masuda, S. Yamada, M. Aizawa, ApJ **822**, 2 (2016)
V. Van Eylen, S. Albrecht, ApJ **808**, 126 (2015)
J.N. Winn, ArXiv e-prints, arXiv:1001.2010 (2010)

Appendix C
Joint Posterior Distributions for the Model Parameters

This appendix shows the corner plots for the joint posterior distributions resulting from the analyses in Chaps. 4 through 6. The two-dimensional and one-dimensional histograms are plotted for selected model parameters to elucidate the nature of the parameter correlations. The inner three contours in the two-dimensional histograms correspond to 1σ, 2σ, and 3σ credible regions of the marginal posteriors. The plots in this appendix are made using `corner.py` by Foreman-Mackey (2016).

C.1 Joint Photometric and Spectroscopic Analysis in Chap. 4

Figures C.1, C.2 and C.3 correspond to the results in Table 4.3 and Fig. 4.9 for the HAT-P-7 system. Figure C.4 shows the result for the Kepler-25 system in Table 4.4 and Fig. 4.10.

C.2 Gravity-Darkened Model Fit in Chap. 5

Figures C.5 and C.6 correspond to the results for Kepler-13Ab in the fourth and sixth columns of Table 5.1: c_2-fitted light-curve solution and joint solution for the B11 stellar parameters. These results are obtained by fitting the Q2 light curve alone, as in Figs. 5.1 and 5.2. Figures C.7 and C.8 correspond to the two joint solutions for HAT-P-7b in Table 5.3.

Fig. C.1 (HAT-P-7) Joint posterior distributions for the most correlated 11 model parameters and ψ for the W09 data set (Table 4.3)

C.3 Joint ETV and Light-Curve Fit in Chap. 6

Figure C.9 corresponds to the joint-fit result in the last column of Table 6.1.

Appendix C: Joint Posterior Distributions for the Model Parameters

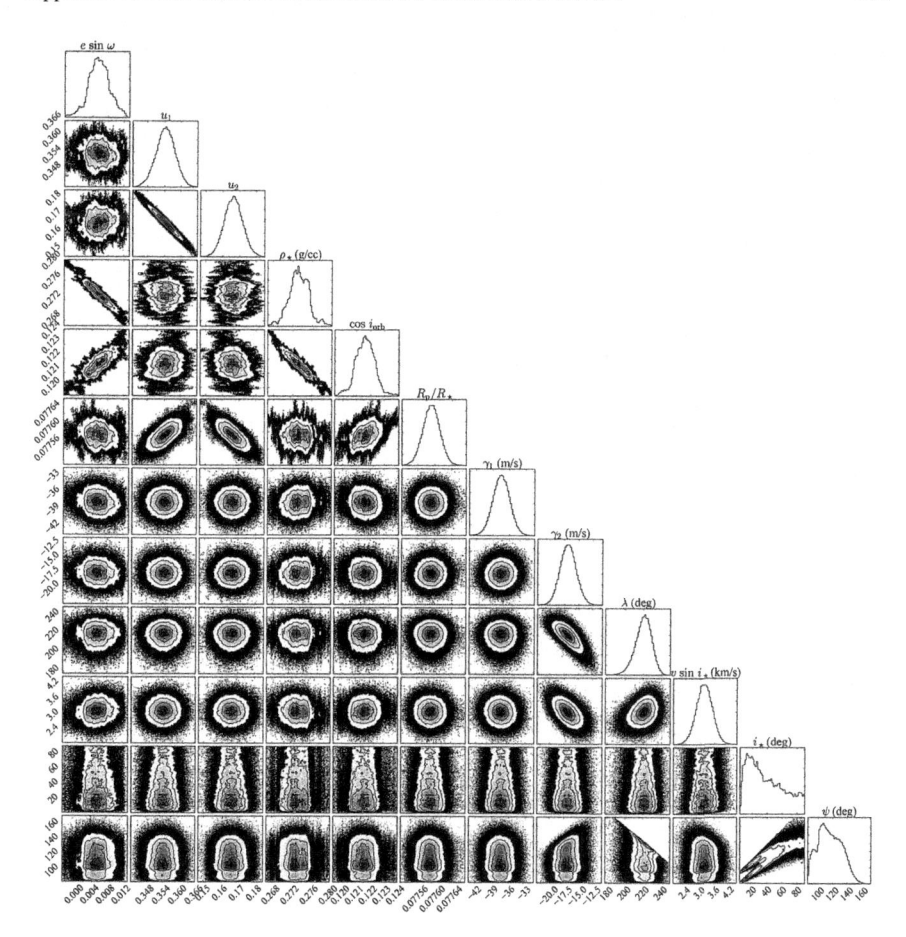

Fig. C.2 (HAT-P-7) Joint posterior distributions for the most correlated 11 model parameters and ψ for the N09 data set (Table 4.3)

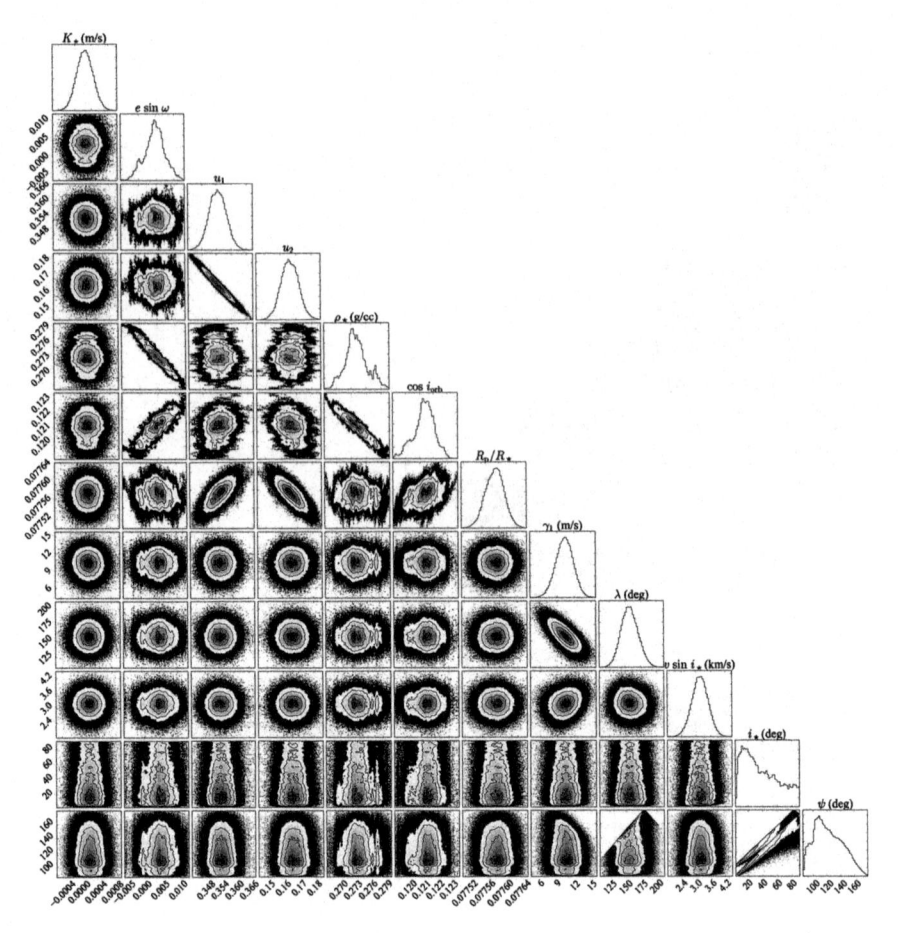

Fig. C.3 (HAT-P-7) Joint posterior distributions for the most correlated 11 model parameters and ψ for the A12 data set (Table 4.3)

Appendix C: Joint Posterior Distributions for the Model Parameters   155

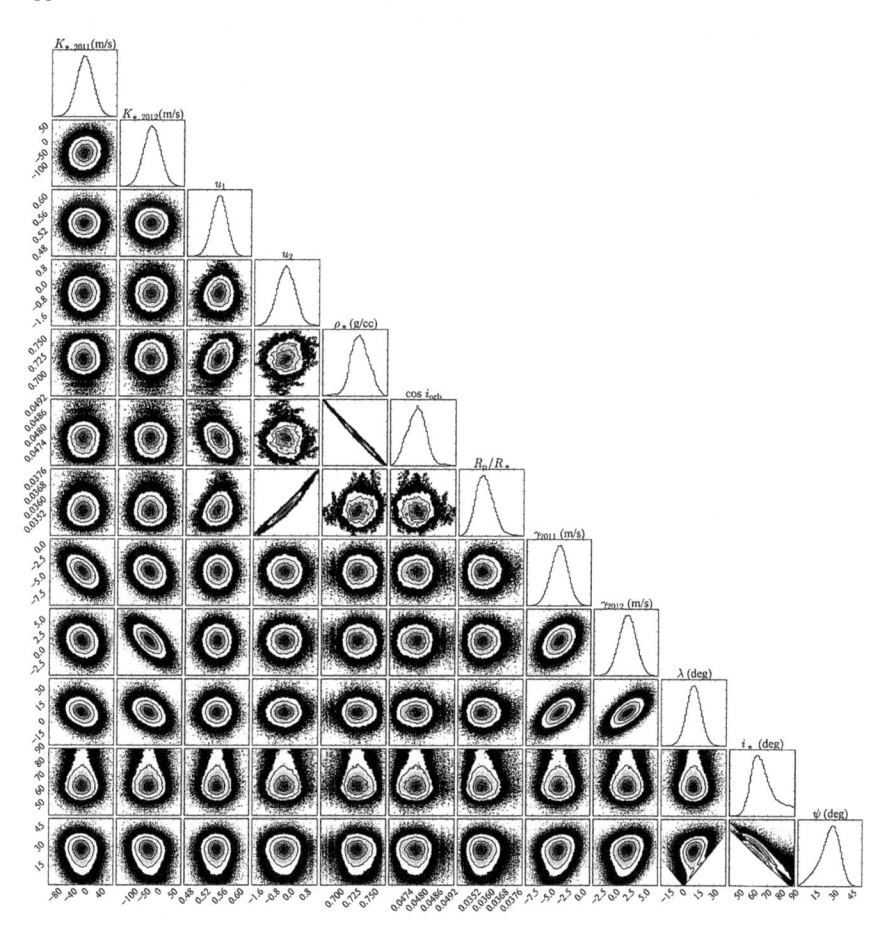

Fig. C.4 (Kepler-25) Joint posterior distributions for the most correlated 11 model parameters and ψ (Table 4.4)

156 Appendix C: Joint Posterior Distributions for the Model Parameters

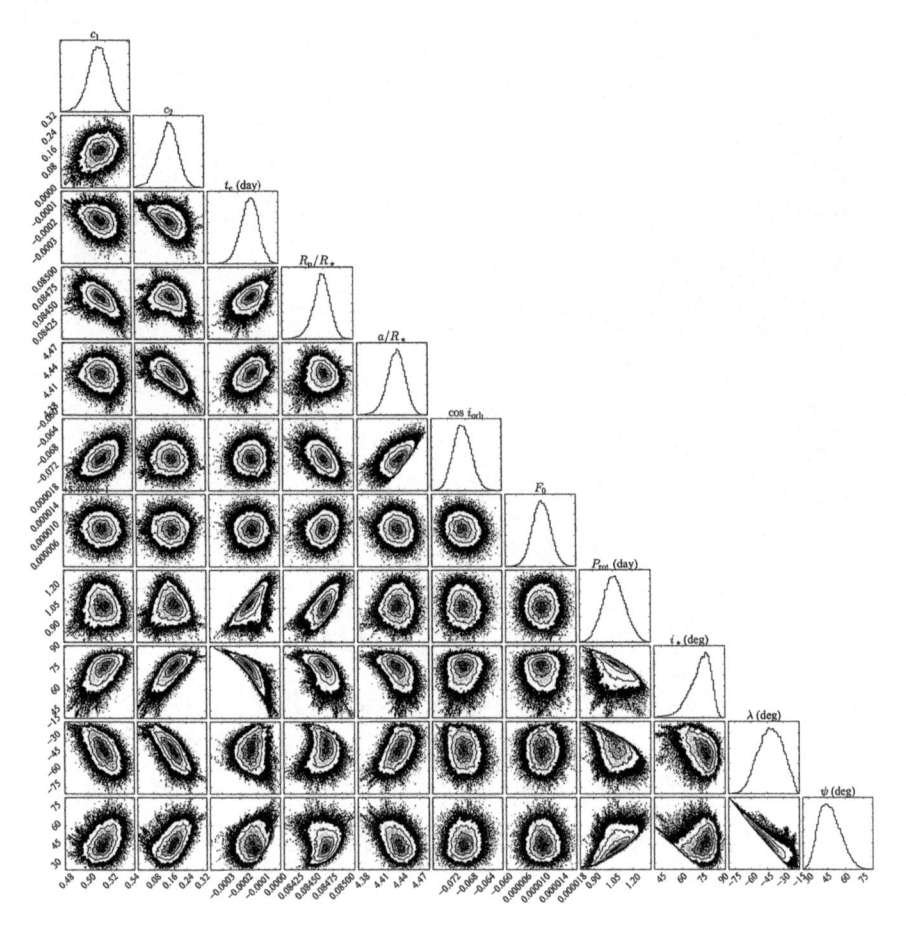

Fig. C.5 (Kepler-13A) Joint posterior distributions for all the model parameters and ψ. This result is for the Q2 transit light curve and adopts B11 set of parameters (Table 5.1, fourth column)

Appendix C: Joint Posterior Distributions for the Model Parameters 157

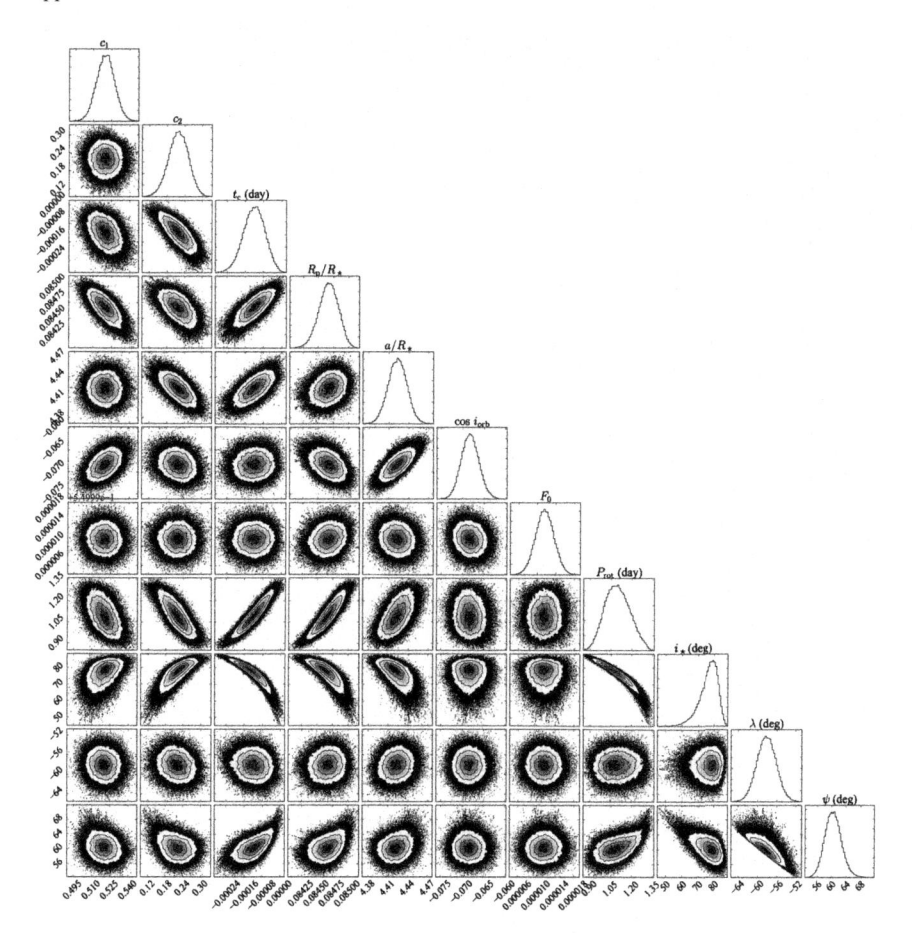

Fig. C.6 (Kepler-13A) Joint posterior distributions for all the model parameters and ψ. This result is for the Q2 transit light curve and adopts B11 set of parameters. Here the spectroscopic constraint on λ is also imposed (Table 5.1, sixth column)

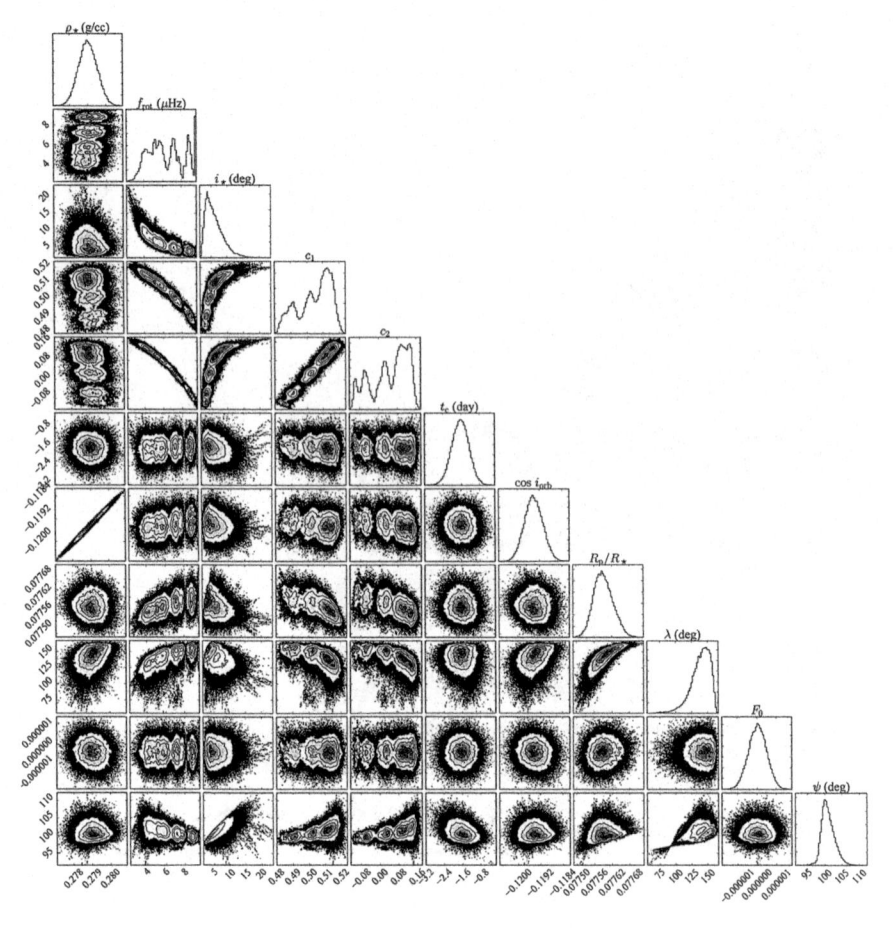

Fig. C.7 (HAT-P-7) Joint posterior distributions for all the model parameters and ψ. This result is for the solution 1 of the joint analysis (Table 5.3)

Appendix C: Joint Posterior Distributions for the Model Parameters 159

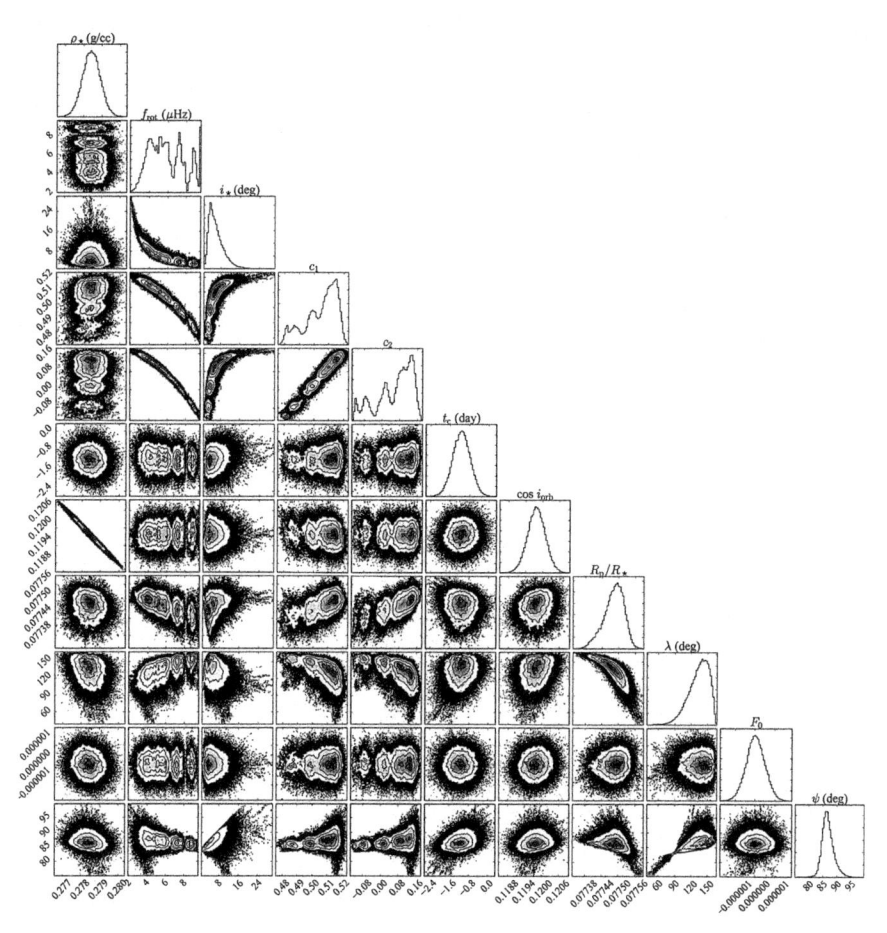

Fig. C.8 (HAT-P-7) Joint posterior distributions for all the model parameters and ψ. This result is for the solution 2 of the joint analysis (Table 5.3)

160   Appendix C: Joint Posterior Distributions for the Model Parameters

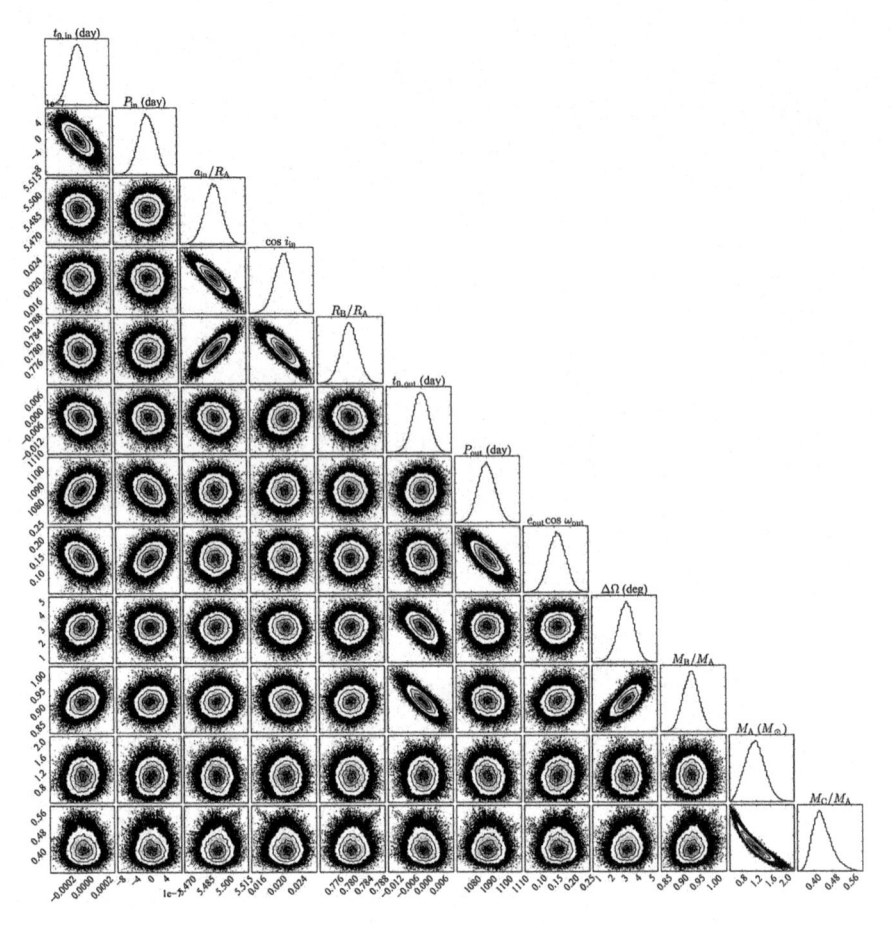

Fig. C.9 Joint posterior distributions for the most correlated 12 model parameters. This result is for the joint analysis that adopts the prior constraint on M_A (Table 6.1, last column)

Reference

D. Foreman-Mackey, J Open Source Softw. **24**, https://doi.org/10.21105/joss.00024 (2016)